CW01236695

Lecture Notes in Electrical Engineering

Volume 487

Board of Series editors

Leopoldo Angrisani, Napoli, Italy
Marco Arteaga, Coyoacán, México
Bijaya Ketan Panigrahi, New Delhi, India
Samarjit Chakraborty, München, Germany
Jiming Chen, Hangzhou, P.R. China
Shanben Chen, Shanghai, China
Tan Kay Chen, Singapore, Singapore
Rüdiger Dillmann, Karlsruhe, Germany
Haibin Duan, Beijing, China
Gianluigi Ferrari, Parma, Italy
Manuel Ferre, Madrid, Spain
Sandra Hirche, München, Germany
Faryar Jabbari, Irvine, USA
Limin Jia, Beijing, China
Janusz Kacprzyk, Warsaw, Poland
Alaa Khamis, New Cairo City, Egypt
Torsten Kroeger, Stanford, USA
Qilian Liang, Arlington, USA
Tan Cher Ming, Singapore, Singapore
Wolfgang Minker, Ulm, Germany
Pradeep Misra, Dayton, USA
Sebastian Möller, Berlin, Germany
Subhas Mukhopadyay, Palmerston North, New Zealand
Cun-Zheng Ning, Tempe, USA
Toyoaki Nishida, Kyoto, Japan
Federica Pascucci, Roma, Italy
Yong Qin, Beijing, China
Gan Woon Seng, Singapore, Singapore
Germano Veiga, Porto, Portugal
Haitao Wu, Beijing, China
Junjie James Zhang, Charlotte, USA

** **Indexing: The books of this series are submitted to ISI Proceedings, EI-Compendex, SCOPUS, MetaPress, Springerlink** **

Lecture Notes in Electrical Engineering (LNEE) is a book series which reports the latest research and developments in Electrical Engineering, namely:

- Communication, Networks, and Information Theory
- Computer Engineering
- Signal, Image, Speech and Information Processing
- Circuits and Systems
- Bioengineering
- Engineering

The audience for the books in LNEE consists of advanced level students, researchers, and industry professionals working at the forefront of their fields. Much like Springer's other Lecture Notes series, LNEE will be distributed through Springer's print and electronic publishing channels.

For general information about this series, comments or suggestions, please use the contact address under "service for this series".

To submit a proposal or request further information, please contact the appropriate Springer Publishing Editors:

Asia:

China, *Jessie Guo, Assistant Editor* (jessie.guo@springer.com) (Engineering)

India, *Swati Meherishi, Senior Editor* (swati.meherishi@springer.com) (Engineering)

Japan, *Takeyuki Yonezawa, Editorial Director* (takeyuki.yonezawa@springer.com) (Physical Sciences & Engineering)

South Korea, *Smith (Ahram) Chae, Associate Editor* (smith.chae@springer.com) (Physical Sciences & Engineering)

Southeast Asia, *Ramesh Premnath, Editor* (ramesh.premnath@springer.com) (Electrical Engineering)

South Asia, *Aninda Bose, Editor* (aninda.bose@springer.com) (Electrical Engineering)

Europe:

Leontina Di Cecco, Editor (Leontina.dicecco@springer.com)
(Applied Sciences and Engineering; Bio-Inspired Robotics, Medical Robotics, Bioengineering; Computational Methods & Models in Science, Medicine and Technology; Soft Computing; Philosophy of Modern Science and Technologies; Mechanical Engineering; Ocean and Naval Engineering; Water Management & Technology)

(christoph.baumann@springer.com)
(Heat and Mass Transfer, Signal Processing and Telecommunications, and Solid and Fluid Mechanics, and Engineering Materials)

North America:

Michael Luby, Editor (michael.luby@springer.com) (Mechanics; Materials)

More information about this series at http://www.springer.com/series/7818

Reji Kumar Pillai · Girish Ghatikar
Ravi Seethapathy · Vijay L. Sonavane
S. A. Khaparde · Pradeep Kumar Yemula
Samir Chaudhuri · Anant Venkateswaran
Editors

ISGW 2017: Compendium of Technical Papers

3rd International Conference and Exhibition on Smart Grids and Smart Cities

Springer

Editors
Reji Kumar Pillai
India Smart Grid Forum (ISGF)
New Delhi, Delhi
India

Girish Ghatikar
Electric Power Research Institute (EPRI)
Berkeley, CA
USA

Ravi Seethapathy
Biosirus
Markham, ON
Canada

Vijay L. Sonavane
Maharashtra Electricity Regulatory
  Commission
Mumbai, Maharashtra
India

S. A. Khaparde
Department of Electrical Engineering
Indian Institute of Technology Bombay
Mumbai, Maharashtra
India

Pradeep Kumar Yemula
Department of Electrical Engineering
Indian Institute of Technology Hyderabad
Hyderabad, Andhra Pradesh
India

Samir Chaudhuri
Working Group 7
India Smart Grid Forum (ISGF)
New Delhi, Delhi
India

Anant Venkateswaran
General Electric (United States)
Aspen, CO
USA

ISSN 1876-1100   ISSN 1876-1119 (electronic)
Lecture Notes in Electrical Engineering
ISBN 978-981-10-8248-1   ISBN 978-981-10-8249-8 (eBook)
https://doi.org/10.1007/978-981-10-8249-8

Library of Congress Control Number: 2018931481

© Springer Nature Singapore Pte Ltd. 2018
This work is subject to copyright. All rights are reserved by the Publisher, whether the whole or part of the material is concerned, specifically the rights of translation, reprinting, reuse of illustrations, recitation, broadcasting, reproduction on microfilms or in any other physical way, and transmission or information storage and retrieval, electronic adaptation, computer software, or by similar or dissimilar methodology now known or hereafter developed.
The use of general descriptive names, registered names, trademarks, service marks, etc. in this publication does not imply, even in the absence of a specific statement, that such names are exempt from the relevant protective laws and regulations and therefore free for general use.
The publisher, the authors and the editors are safe to assume that the advice and information in this book are believed to be true and accurate at the date of publication. Neither the publisher nor the authors or the editors give a warranty, express or implied, with respect to the material contained herein or for any errors or omissions that may have been made. The publisher remains neutral with regard to jurisdictional claims in published maps and institutional affiliations.

Printed on acid-free paper

This Springer imprint is published by Springer Nature
The registered company is Springer Nature Singapore Pte Ltd.
The registered company address is: 152 Beach Road, #21-01/04 Gateway East, Singapore 189721, Singapore

Contents

Part I Smart Grid and IoT Technologies

Developing SMARTGRID Projects with Global
Perspective in Indian Scenario 3
Amit R. Kulkarni, M. S. Ballal and Shrikant S. Rajurkar

Smart Energy Metering Using LPWAN IoT Technology 19
Shobhit Jain, M. Pradish, A. Paventhan, M. Saravanan and Arindam Das

Smart Grid Technologies: Distribution Automation,
Microgrids, and Cyber Security 29
S. R. Vijayan

Impact Analysis of Attacks Using Agent-Based
SCADA Testbed 41
M. Siddharth Rao, Rajesh Kalluri, R. K. Senthil Kumar,
G. L. Ganga Prasad and B. S. Bindhumadhava

Power Exchange and Its Significance to Enhance
the Deployment of Smart Microgrids in India and Key
Barriers in Its Adoption 55
Furkan Ahmad, Mohammad Saad Alam, Reena Suri, Akhilesh Awasthy
and M. Shahidehpour

Low-Cost Spark-/Arc-Free Retrofit Smart Grid Switches
Improve Distribution Quality and Reduce Distribution
Losses Substantially 69
G. V. Sukumara and Vijay L. Sonavane

Micro-phasor Measurement Units (μPMUs) and Its
Applications in Smart Distribution Systems 81
Alok Jain and Suman Bhullar

A Rule-Driven Architecture to Address Interoperability in an IEC 61850 Series-Based Power Utility Automation System 93
Mayank Sharma and Thomas Rudolph

Software Defined Networking for Smart Grid Communications and Security Challenges 103
M. U. Shaileshwari, K. S. Nandini Prasad and A. Paventhan

Digital Utility 113
Deepak Chaudhary

Analysis of Communication Channel Attacks on Control Systems—SCADA in Power Sector 115
Rajesh Kalluri, Lagineni Mahendra, R. K. Senthil Kumar, G. L. Ganga Prasad and B. S. Bindhumadhava

Plug and Operate Power and Distribution Transformer Technology 133
Deepal Shah

Part II E-Mobility

Public Opinion on Viability of xEVs in India 139
Mohd. Saqib, Md. Muzakkir Hussain, Mohammad Saad Alam, M. M. Sufyan Beg and Amol Sawant

Technical and Economic Feasibility Analysis for deployment of xEV Wireless Charging Infrastructure in India 151
Vatsala, Raqib Hasan Khan, Yash Varshney, Aqueel Ahmad, Mohammad Saad Alam and Rakan C. Chaban

Viability of xEVs in India: A Public Opinion Survey 165
Mohammad Asaad, Prashant Shrivastava, Mohammad Saad Alam, Yasser Rafat and Reji Kumar Pillai

Thermal Management Solutions of Lithium-Ion Energy Storage Batteries for xEV Deployment in North India 179
Mohd Yaqzan, Yasser Rafat, Sheikh Abdullah and Mohammad Saad Alam

Electric Vehicle Charging Infrastructure in India: Viability Analysis 193
Wajahat Khan, Furkan Ahmad, Aqueel Ahmad, Mohammad Saad Alam and Akshay Ahuja

Part III Renewable Energy, Microgrids and Energy Storage

Challenges in Implementation of Virtual Synchronous Generator 209
Ganesh N. Jadhav, Sadik J. Shaikh and Omkar N. Buwa

Efficiency Gain Using DC Microgrid and BLDC Machine-Based 48 V Air Cooler................................. 223
Sriram Narayanamurthy, Pradheep Ganesan, Ashok Jhunjhunwala and Prabhjot Kaur

Integrating Energy Efficiency with Renewables and Energy Storage for a Smarter and Greener Residential Solution............ 231
Satish Kumar, Smita Chandiwala, Vaibhav Rai Khare and Manish Pant

Smart Solutions and Opportunities for Key Challenges in Renewables Integration and Electric Vehicles Integration to a Conventional Grid.. 237
Goutham Yelmanchli

Solar Agricultural Pump Without Electric Motor................. 251
G. V. Sukumara and Vijay L. Sonavane

Applicability of Error Limit in Forecasting and Scheduling of Wind and Solar Power in India............................. 263
Abhik Kumar Das

Smart Microgrids: Re-visioning Smart Grid and Smart City Development in India.. 273
Larisa Dobriansky, Girish Ghatikar and Dan Ton

Part I
Smart Grid and IoT Technologies

Developing SMARTGRID Projects with Global Perspective in Indian Scenario

Amit R. Kulkarni, M. S. Ballal and Shrikant S. Rajurkar

Abstract This paper discusses in brief various aspects related to SMARTGRID technology developments and their importance in adopting in Indian scenario. It covers some of the practical SMARTGRID technologies developmental aspects in the state of Maharashtra in Indian context. It also discusses initiatives taken by MAHATRANSCO in developing these technologies in the field of Wide Area Measurement System (WAMS). It covers some of the aspects like understanding grid disturbances/events in the system with the help of WAMS to know about oscillations in the grid during these events along with case studies. It discusses in brief improving situational awareness of the grid based on WAMS. This paper covers consideration to Flexible AC Transmission Systems (FACTS) devices in the system, as a part of initial steps taken by MAHATRANSCO in understanding dynamic compensation-related requirements in the system to meet dynamic voltage variations. It also covers importance of Dynamic Line Rating (DLR) technology for transmission utilities in Indian scenario. It further focuses on how to streamline developmental activities amongst centre and state as far as some of the aspects of Renewable Energy utilisation and integration into the grid and Renewable Energy Management Systems (REMS)/Renewable Energy Management Centre (REMC) in India at centre and state Level are concerned.

Keywords WAMS · SMARTGRID · Situational awareness · Oscillations
FACTS · DLR · REMS/REMC

A. R. Kulkarni (✉) · S. S. Rajurkar
MAHATRANSCO, Mumbai, India
e-mail: amk4.ind@gmail.com

S. S. Rajurkar
e-mail: ssrajurkar@gmail.com

M. S. Ballal
VNIT, Nagpur, India

© Springer Nature Singapore Pte Ltd. 2018
R. K. Pillai et al. (eds.), *ISGW 2017: Compendium of Technical Papers*, Lecture Notes in Electrical Engineering 487, https://doi.org/10.1007/978-981-10-8249-8_1

1 Introduction

This paper covers six sections. Section 2 gives brief description of MAHATRANSCO initiatives in WAMS development; Sect. 3 discusses utilisation of WAMS for analysing the grid events along with case studies. It also covers use of Prony analysis for understanding oscillations in the system during these events and also focuses on utilisation of situational awareness system. Section 4 describes other SMARTGRID technology developmental aspects in MAHATRANSCO. Section 5 presents various issues pertaining to Renewable Energy (RE) utilisation, its integration and related important aspects needing attention in Indian scenario.

Figure 1 shows 400 kV network overview of MAHATRANSCO system. MAHATRANSCO is the largest State Transmission Utility (STU) in India with 634

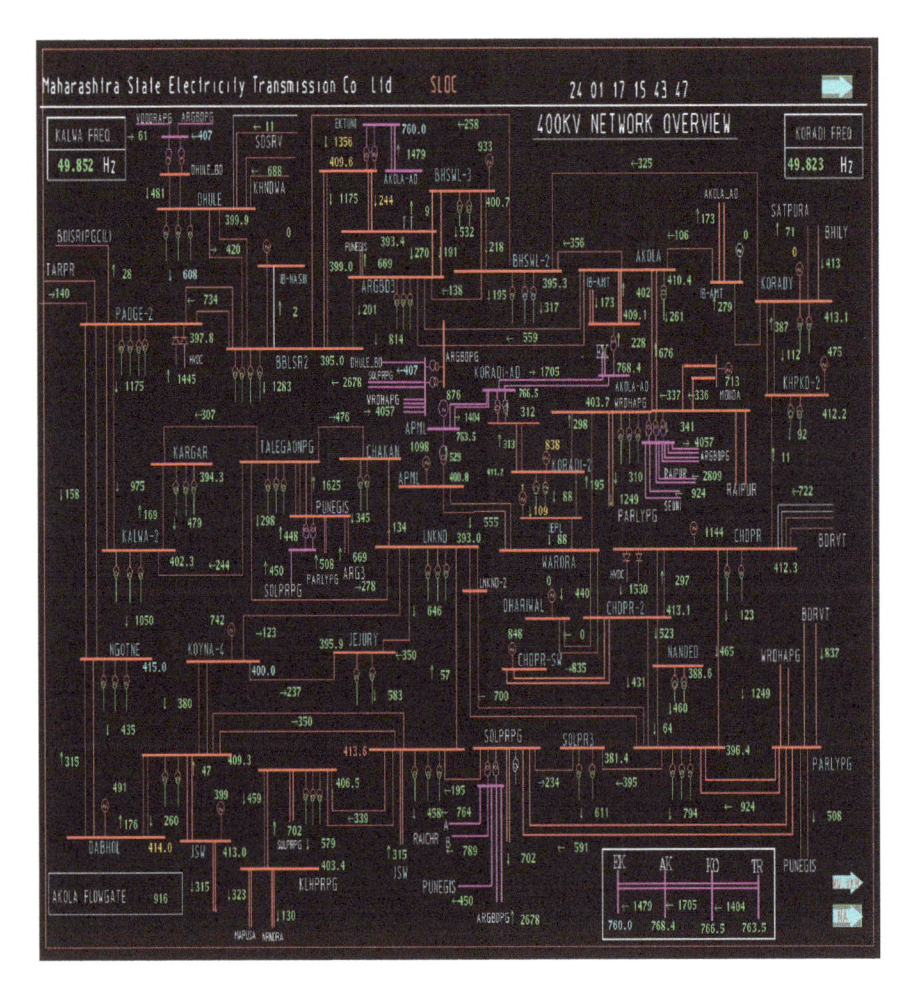

Fig. 1 400 kV network overview of MAHATRANSCO

EHV substations, 110,000 MVA of transformation capacity and 47,000 circuit kilometres of EHV transmission lines.

This paper discusses in brief initiatives taken by MAHATRANSCO in Indian context towards development of some of the SMARTGRID technologies along with case studies for the same. This paper also covers in brief, issues needing attention while harnessing RE power in Indian context.

2 MAHATRANSCO Initiatives in Development of Wide Area Measurement System

Figure 2 shows architecture of WAMS infrastructure in MAHATRANSCO system. Present MSETCL's Wide Area Measurement System (WAMS) infrastructure includes Phasor Measurement Units (PMUs), GPS clock and antenna installed at various locations at twelve 400 kV and three 220 kV level substations. Central Phasor Data Concentrator (PDC), visualisation software and historian software are installed at State Load Despatch Centre (SLDC) of Maharashtra located at Kalwa. The Real time data transfer of parameters like Voltages (V), Currents (I), Active Power (P), Reactive Power (Q), Frequency (F), Rate of Change of frequency

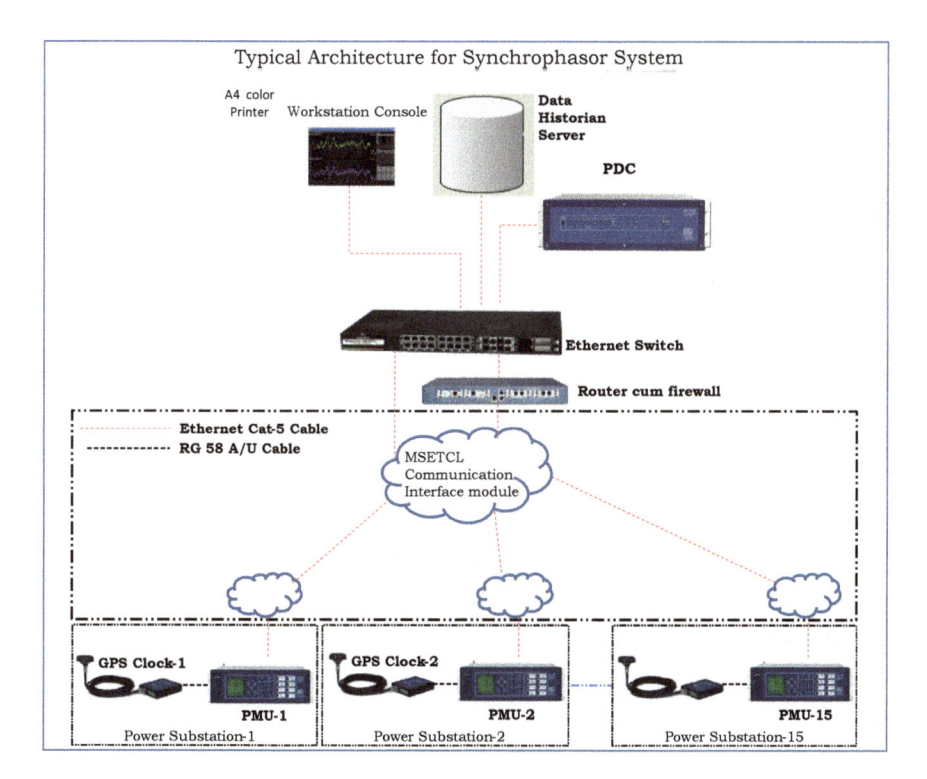

Fig. 2 Architecture of WAMS infrastructure in MAHATRANSCO

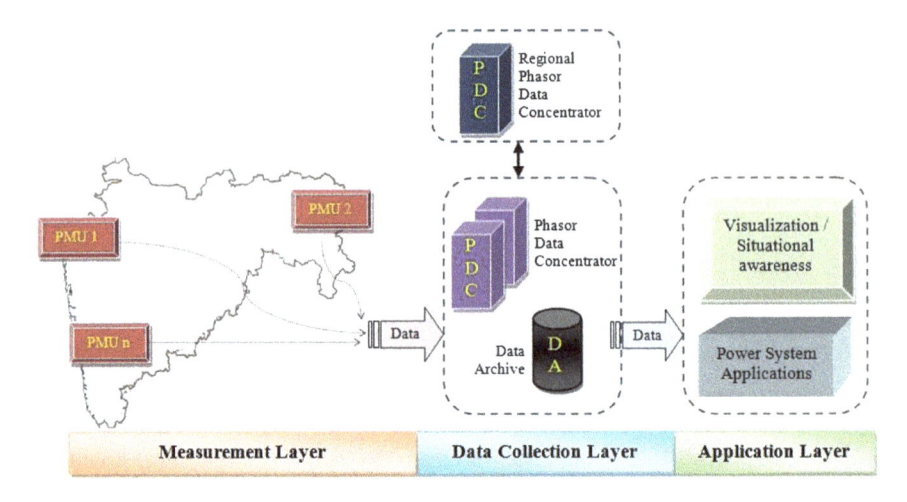

Fig. 3 Three-layer architecture of WAMS infrastructure in MAHATRANSCO

(DF/DT) and Delta (δ) is done from PMUs located at important locations and Critical substations in MAHATRANSCO grid to PDC located at State Load Despatch Centre (SLDC)-Kalwa.

Figure 3 shows three-layer architecture showing how the data flows from physical WAMS infrastructure spread across MAHATRANSCO system in Maharashtra state in India to data storage/archiving system comprising of Phasor Data Concentrator (PDC), historian, etc., and further to application software.

3 Utilisation of Synchrophasor System for Grid Event Analysis and for Detecting Oscillations

This section discusses utilisation of synchrophasor system for analysing grid events with case studies for the same. It also discusses in brief Prony analysis technique used along with PMU data to understand type of oscillatory modes excited during these events.

3.1 Case Study-1: CGPL Ultra Mega Power Plant (UMPP) Blackout Analysis Dated on 13 July 2016 in Western Regional Grid in India Using Synchrophasor System

On 12 July 2016 due to insulator tracking at Varsana substation and subsequent fault, there was loss of all 400 kV elements from 400/220 kV Varsana substation in

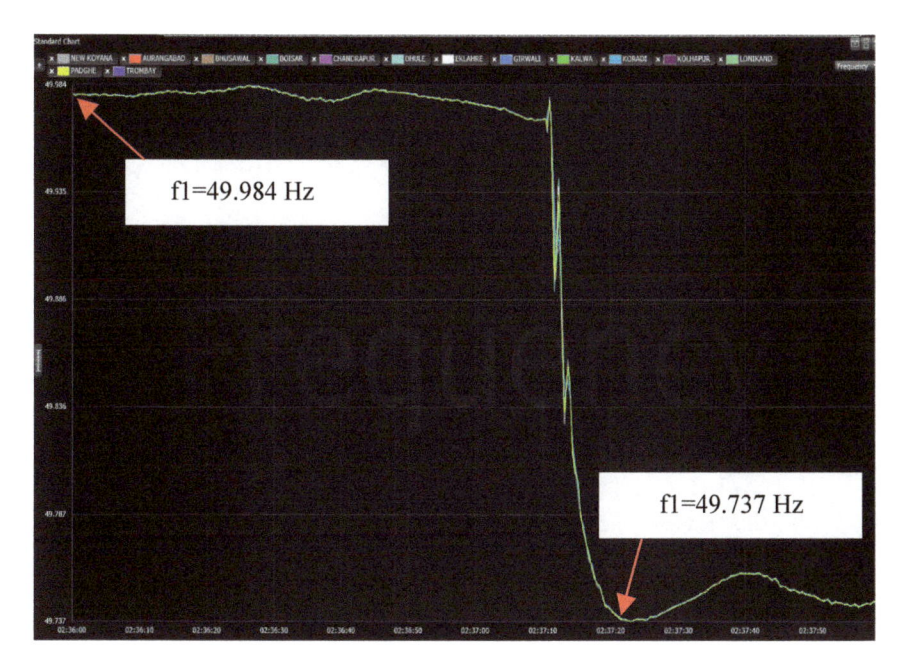

Fig. 4 Frequency behaviour during CGPL-UMPP occurrence

Gujarat. On 13 July 2016 around 2:37 am, 400 kV Bachau–Varsana line was charged from Bachau end. Seven seconds after charging this line, fault was developed in Y-phase which later got converted into three-phase fault at gantry end of this line at Varsana end. It was seen that all the remaining lines from CGPL UMPP along with 400 kV Bachau–Ranchodpura D/C, 220 Bachau–Morbi tripped. This resulted in blackout of CGPL UMPP with tripping of all its running Units, i.e. Unit-10 [760 MW], Unit-30 [742 MW], Unit-40 [757 MW], Unit-50 [619 MW], tripped with total generation loss of 2878 MW. The loss of generation resulted in frequency decay further resulting in df/dt-based load shedding operation to the tune of around 119 MW in GETCO system in Gujarat. Figure 4 shows frequency behaviour as sensed by PMUs in MAHATRANSCO grid during CGPL-UMPP disturbance. Figure 5 shows df/dt as observed by PMUs at 220 kV Boisar and 220 kV Eklahare substations in MAHATRANSCO grid during this incidence. The df/dt as indicated by PMUs was found to be around −0.3212 Hz/s.

In order to understand type of oscillatory modes excited during CGPL-UMPP occurrence dated on 13 July 2016, Prony analysis was carried out with frequency measured from PMUs at 220 kV Eklahare and 400 kV Padghe substations in MAHATRANSCO grid. Figures 6 and 7, respectively, show plot for Prony approximate and PMU measured frequency response for 220 kV Eklahare and 400 kV Padghe substation in MAHATRANSCO grid.

Table 1 shows some of the important results of Prony analysis indicating amplitude, damping and frequency of oscillations during CGPL-UMPP

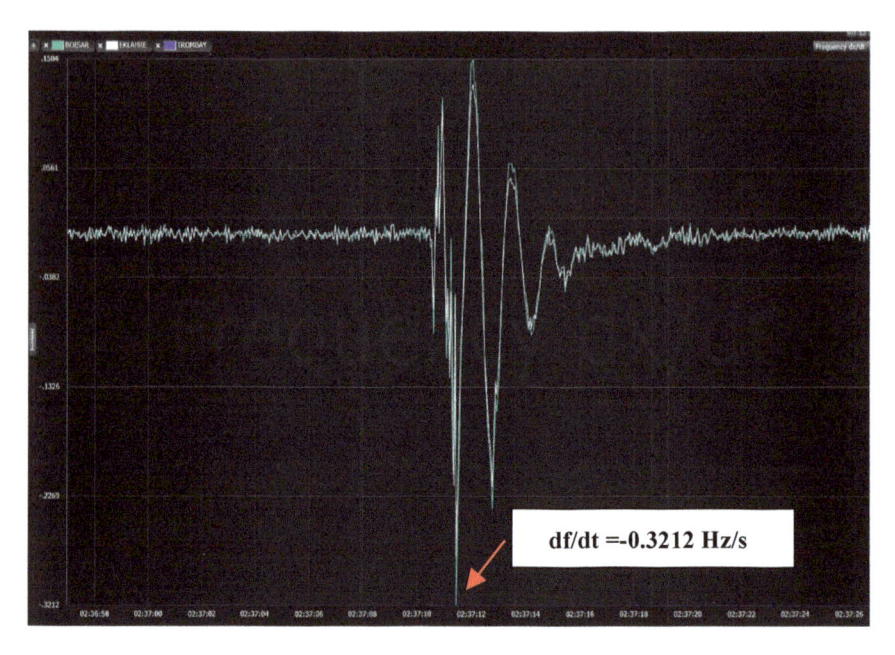

Fig. 5 d*f*/d*t* as observed by PMUs at 220 kV Boisar and Eklahare substations during CGPL-UMPP occurrence

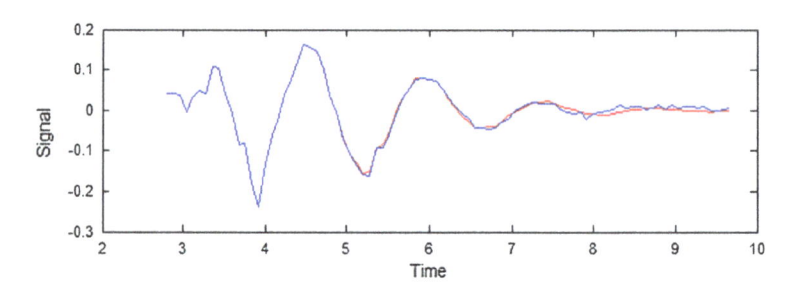

Fig. 6 Plot for Prony approximate and PMU measured frequency response for 220 kV Eklahare substation in MAHATRANSCO system

disturbance. From this Table 1 based on results of Prony analysis of PMU data for 220 kV Eklahare substation, 400 kV Padghe and other substations where PMUs are placed in MAHATRANSCO system, it can be said that inter-area modes of oscillations of frequency around 0.55 Hz were prominently seen along with negative damping during CGPL-UMPP disturbance.

Figure 8 indicates angular separation between 400 kV Padghe and Aurangabad substations in MAHATRANSCO grid before CGPL-UMPP occurrence, whereas Fig. 9 indicates angular separation between 400 kV Padghe and Aurangabad substations in MAHATRANSCO grid after CGPL-UMPP occurrence. This highlights

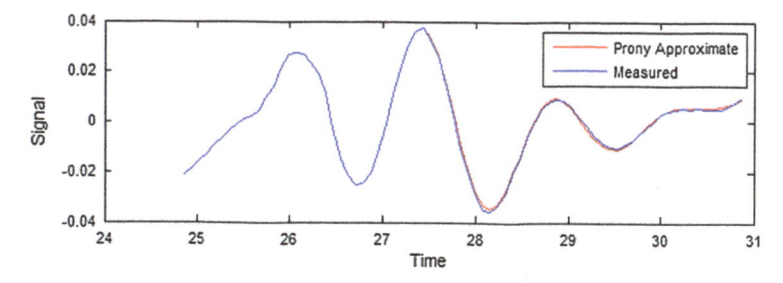

Fig. 7 Plot for Prony approximate and PMU measured frequency response for 400 kV Padghe substation in MAHATRANSCO system

Table 1 Results of Prony analysis for case study-1

Sr. no.	PMU location	Amplitude	Damping	Freq. (Hz)
1	400 kV Padghe	0.083	−2.2	0.91
		0.066	−0.43	0.78
		0.037	−0.34	0.57
		0.01	−0.045	0.23
2	220 kV Eklahare	1.0	−0.86	0.73
		1.0	−1.2	0.55

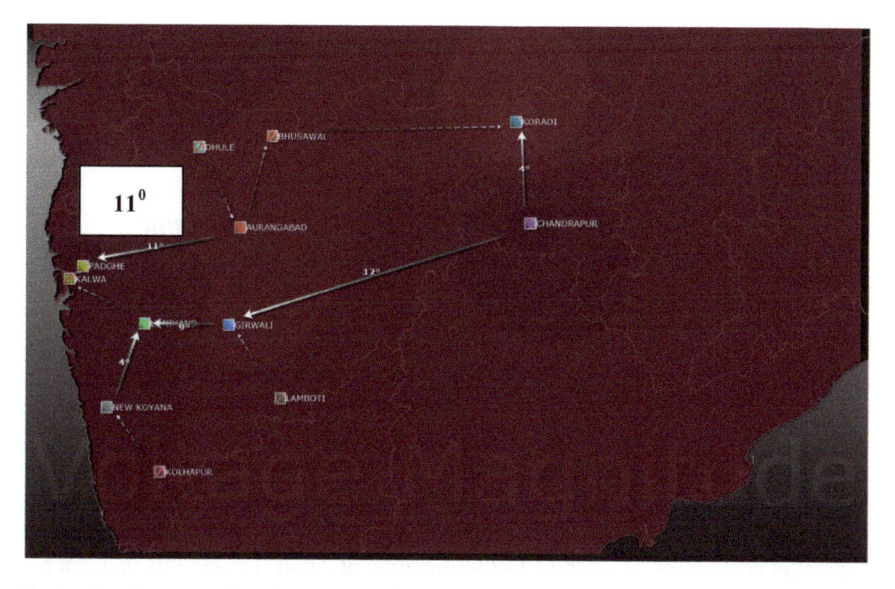

Fig. 8 Angular separation between 400 kV Padghe and Aurangabad before CGPL-UMPP occurrence

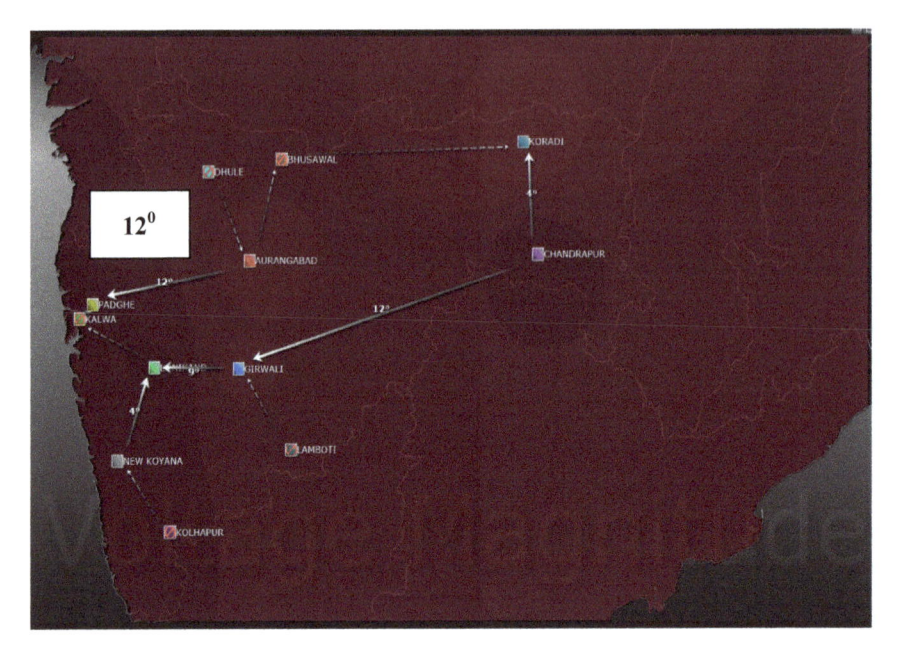

Fig. 9 Angular separation between 400 kV Padghe and Aurangabad after CGPL-UMPP occurrence

the fact that even small angular difference of 1° between two nodes can be detected using WAMS. This assists in improving situational awareness of the grid during disturbances which can further be used to devise requisite protection and control actions as per system requirements to improve the system stability.

3.2 Case Study-2: Analysis of HVDC Spikes Phenomena Dated on 07 October 2015 in Western Regional Grid in India Using Synchrophasor System

MAHATRANSCO is the only State Transmission Utility (STU) in India which owns and maintains ±500 kV bipolar HVDC link between Chandrapur in Eastern and Padghe in Western part of Maharashtra having capacity to transfer 1500 MW. On 07 October 2015, pole-1 was on outage since 7:56 am. Pole-2 was operating in monopolar ground return mode in which full return current flows through electrode line connected to earth electrode station at Anjur. The electrode line comprises of two parallel conductors EL-1 and EL-2 terminating at Anjur. During this incidence, it was found that one of the conductors of EL-1 was burnt and was hanging near to Padghe. As conductor was not completely snapped, high impedance path resulted in uneven sharing of current between the two conductors further activating 'Electrode

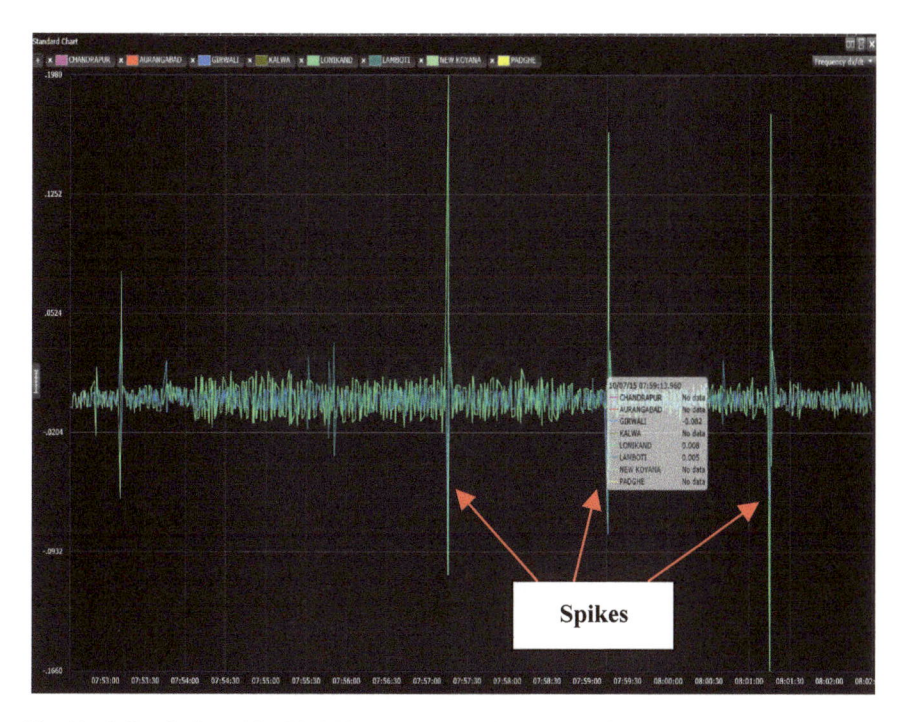

Developing SMARTGRID Projects with Global Perspective ... 11

Fig. 10 Spikes indicated by WAMS during events dated on 07 October 2015

Unbalance Protection'. It was seen that trip command got reset after every two minutes by timer. This was indicated by spikes in the system parameters; one of them is for frequency as shown in Fig. 10. These repetitive natured spikes observed exactly at every 2 min on synchrophasor system. This phenomenon of spikes stopped only after ramping down power carried by pole-2 from 750 to 600 MW around 8.50 am.

This phenomenon was analysed with the help of Prony analysis to understand kind of oscillations seen in the system during this incidence. Figure 11 indicates frequency signal as obtained from PMU at 220 kV Boisar substation in MAHATRANSCO grid considered for analysis purpose. In this, section 'a' of signal indicates first spike and section 'b' indicates another spike as observed during this event.

Figures 12 and 13 show comparison of Prony approximate and PMU measured frequency signal for section 'a' and section 'b' of Fig. 11, respectively.

Table 2 gives some of the important results of Prony analysis. During this analysis, it was observed that these oscillations were present during duration of this event and disappeared after power order for pole-2 was reduced from 750 to 600 MW. From this, it can be said that control mode of around 4.7 Hz was seen with negative damping during this incidence. This can be related to control system

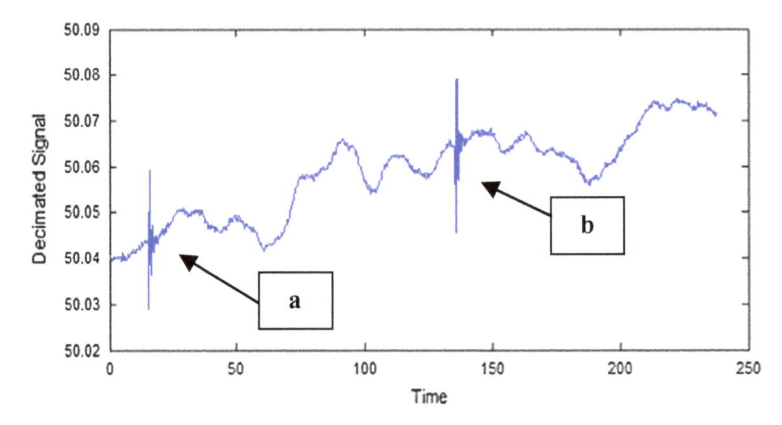

Fig. 11 Frequency signal as indicated by PMU of 220 kV Boisar considered for Prony analysis

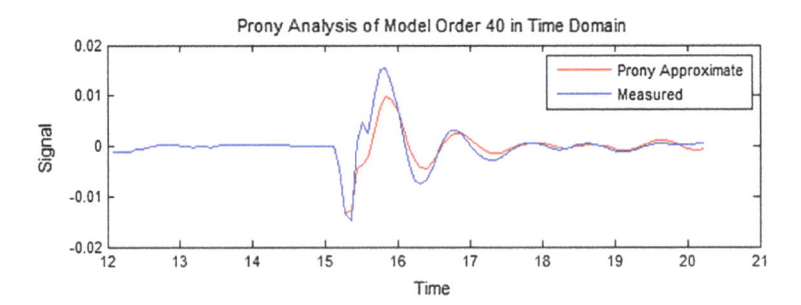

Fig. 12 Plot for Prony approximate and PMU measured frequency response for 220 kV Boisar substation at section 'a'

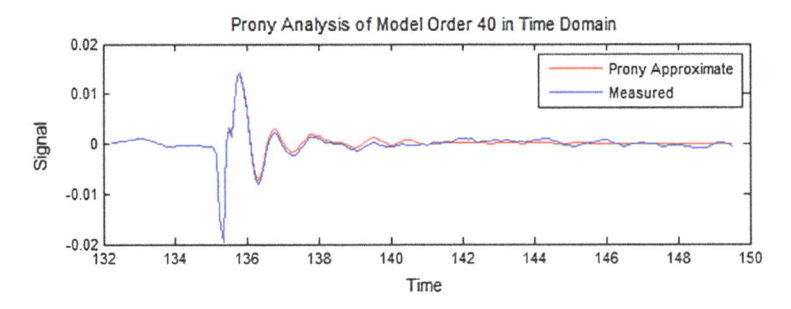

Fig. 13 Plot for Prony approximate and PMU measured frequency response for 220 kV Boisar substation at section 'b'

Table 2 Results of Prony analysis for case study-2

Sr. no.	PMU location	Amplitude	Damping	Freq. (Hz)
1	220 kV Boisar (Instant 'a' in Fig. 11)	0.028	−1.1	3.2
		0.028	−1.2	3.5
		0.025	−1.3	4.1
2	220 kV Boisar (Instant 'b' in Fig. 11)	0.021	−1.6	4.2
		0.065	−2.5	4.3
		0.011	−1.6	4.7

behaviour of HVDC system which was activated after 'Electrode Unbalance Protection' initiation during this event.

4 Other SMARTGRID Initiatives in MAHATRANSCO

Apart from WAMS, MAHATRANSCO system adopts High Tension Low Sag (HTLS) technology, IEC-61850 compliant numerical relaying system, IEC-104-based SCADA system, ABT metering system for effective energy accounting, ERP system for streamline works of different verticals of the organisation.

4.1 Adoption to FACTS Technology

MAHATRANSCO system experiences dynamic voltage variations in many parts of the system having RE concentration like that observed in areas like Karad, Lamboti, Dhule, Nashik. In view of this, MAHATRANSCO is also planning to adopt other SMARTGRID technology options like adoption to Flexible AC Transmission System (FACTS) devices to cope up with multiple challenges of reactive power management, to enhance system transmission capacity and improve system stability. Experience in this field world over has shown that overall costs and performance of power system operation can be best managed if voltage control and VAR control are well-integrated. In one such exercise being carried out in MAHATRANSCO, studies for adoption of dynamic compensation for Renewable Energy (RE) rich Karad area in Maharashtra is undertaken. The preliminary studies indicate that, provision of FACTS devices like Static VAR Compensators (SVC), Static Synchronous Compensator (STATCOM) can be considered to address the issue of varying voltages beyond acceptable limits in this area.

4.2 Adoption to Dynamic Line Rating (DLR) Technique

In Maharashtra's over 40,000 MW of installed generation capacity, RE share is of around 17% and with State and Central Government's RE centric policies it is posed to increase in coming years. In this context, Karad area of MAHATRANSCO grid in the state of Maharashtra having around 2500 MW of wind generation, multiple co-generation plants injecting into the grid along with solar generation or Nashik and Ahmednagar area where high wind concentration exists are suitable locations to adopt DLR technology. This technology option thus can be adopted in above-mentioned areas where high renewable penetration like wind energy exists. Line loading can be changed dynamically considering weather conditions by adopting Dynamic Line Rating (DLR) in such cases. Addition of new lines to relieve the congestion during peak RE periods is not an economical solution, as high wind periods are limited throughout the year and the same is true for solar peaks. In such scenario, DLR offers a more desirable solution as high wind speeds also mean increased capacity of lines if proper monitoring of temperature and corresponding line flow increase is coordinated. Generally, it is observed that operator at load dispatch centres loads the line conservatively and there exists a scope to transfer more power through the same corridor, so DLR technology can be used at suitable locations for enhancing power transfer capability. DLR technique can be used in association with PMUs to devise Remedial Action Schemes (RAS) to enhance reliability of the system under uncertain, intermittent Renewable Energy (RE) sources in RE-rich areas in the state of Maharashtra in India.

5 Managing the Renewable Energy

In India during last one decade, it was observed that, in order to utilise renewable sources of energy effectively and to reduce dependence on imported fuel, Central and various State Governments are promoting renewable sources of energy with policies like Renewable Purchase Obligation (RPO) and with various regulations in this regard. With launch of Jawaharlal Nehru National Solar Mission (JNNSM), there is a rise in utilisation of solar energy in India. Maharashtra being one of the leading state in India harnessing Renewable Energy (RE), it has installed capacity of around 7000 MW of RE generation. Under Government of Maharashtra's new RE policy, it is proposed to add 14,400 MW of RE in coming years, thus taking the total to approximately 21,100 MW. Out of this, 7500 MW would come from solar, 5000 MW from wind and remaining 1900 MW from small hydro, baggase-based co-generation, biomass, Municipal Solid Waste (MSW), industrial waste, etc. Considering uncertainty, intermittency and variability of RE sources like wind and solar, their integration to the grid for effective utilisation is a big challenge, which can be dealt with by adopting Renewable Energy Management Systems (REMS).

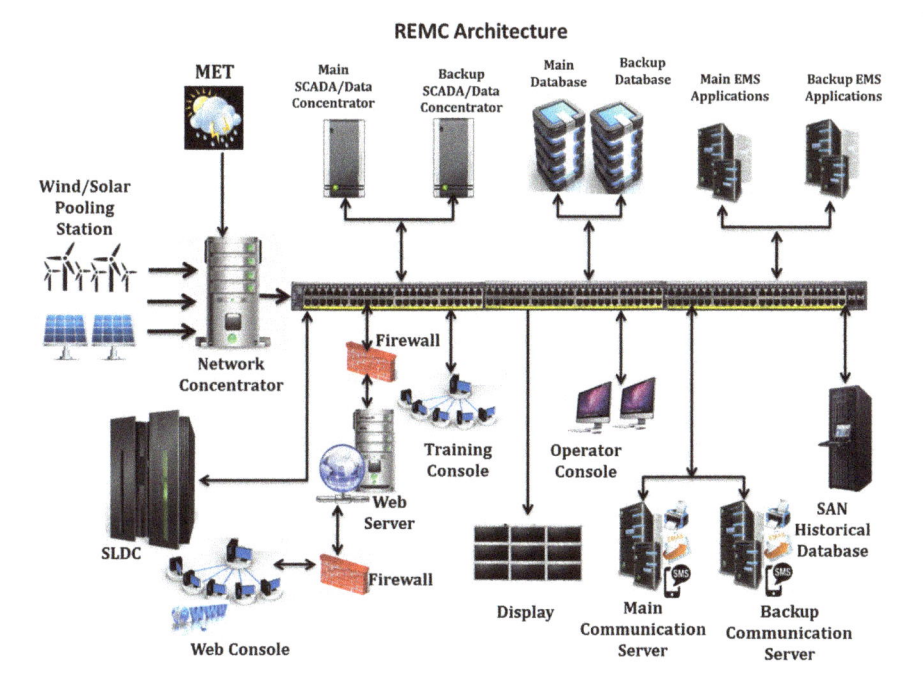

Fig. 14 Typical architecture of REMC

Figure 14 shows typical architecture of REMC that can be made operational for better grid management.

For real-time operation, monitoring and control of RE sources and for its effective integration, RE sources like wind must possess advanced characteristics like fault ride through capability, VAR control and regulation, active power control, ramping and curtailment, primary frequency regulation, inertial control, short-circuit current control. Whereas consideration to Solar irradiance, effect of cloud on solar output, solar energy prediction etc., are some of the important factors for real time operation, monitoring and control of solar energy. Taking into account the importance of RE integration and data transfer from RE generation clusters to central dispatch centre data transfer architecture can be modified as per requirements.

5.1 Issues Needing Attention for Utilisation and Integration of Renewable Energy in Indian Context

(1) Operationalising REMS/REMC infrastructure as per road map overcoming the challenges in its implementation.

(2) As most of the Wind Turbine Generators (WTGs) in India don't have FRT features, motivating them to adopt newer technologies compliant to this is a challenge.
(3) Adopting regulatory and policy frameworks in line with latest developments in RE field and streamlining centre and state policies in this regard.
(4) Establishment of communication infrastructure, weather stations, RTUs, SCADA system in timely manner in RE-rich states for reliable real-time data transfer to central location and funding provisions for the same.
(5) Coping with the challenges of grid management with increased RE penetration.
(6) Coping up with the challenges of operating the RE systems with hourly, subhourly and 15 min block under Availability-Based Tariff (ABT) system in Indian context.
(7) Achieving accuracy in RE forecasting, load forecasting and real-time scheduling of RE power.
(8) Understanding and addressing impacts of complex interactions between four major stakeholders in RE sector such as RE developer focusing on maximising profits, grid operators starving for secure, reliable and economic operations of the grid, utilities trying to opt for RE power at competitive rates and financial traders is an important issue to deal with.
(9) Issuing guidelines regarding RE and SCADA applications covering various aspects related to power plants, turbines, substation and load despatch level.
(10) Plan for developing ensemble forecasting and provision of funding for the same.
(11) Studies related to understanding impacts of dynamic behaviour of RE sources on grid operation.
(12) Developing economical and effective ancillary services support, energy storage systems.
(13) Guidelines regarding protection system requirements of RE sources.
(14) Addressing reactive power management issues with high level of RE penetration and provision of dynamic compensation devices for the same.
(15) Efficient scrutiny, checking of Detail Project Reports (DPRs) of RE developers and standardised mechanism for approving the same along with developing mechanism for periodic RE project review.
(16) Ensuring timely completion of RE projects with due telemetry, protection system, smart metering system, SCADA system, forecasting system at developer end as well as required reactive power compensation provided as per system requirement.
(17) Developing effective mechanism for RE pricing as well as for incentives to developer, utilities, consumers, etc.
(18) Creating environment in which RE sources can be effectively used with ancillary services support, energy storage systems, demand response initiatives and with electrical vehicles.

(19) Human resource development with due training to execute RE projects and to deal with various technical issues while operationalising REMC and real-time grid operations.
(20) In REMC proposed at state level in India forecasting tool must give output of state-level aggregated RE forecast, so for this internationally best proven methodology used by utilities needs to be adopted.

References

1. www.powermin.nic.in
2. www.cercind.gov.in

Smart Energy Metering Using LPWAN IoT Technology

Shobhit Jain, M. Pradish, A. Paventhan, M. Saravanan
and Arindam Das

Abstract The last-mile networking for Internet of Things (IoT) applications using short-range networks in ISM band such as IEEE 802.15.4 LoWPAN mesh, WiFi, Zigbee, Bluetooth Low Energy has been studied widely in the last few years with demonstration in many industrial scenarios. However, the reliable connectivity in last-mile scenarios like individual energy meter in the home area network (HAN) connecting to the data concentrator in turn to the meter data management systems (MDMS) through WAN connectivity is considered to be a challenge in certain areas. There are emerging low-power WAN (LPWAN) technologies such as LoRa, Wi-SUN, Sigfox—all operating in unlicensed band, and NB-IoT—in licensed band that can provide alternative long-range connectivity option for realizing IoT networks. In this paper, we discuss the Indian smart metering deployment in both rural and urban scenarios where the short-range IoT solution built-in may not always work best to the needs of long-range expectations. Further, we highlight how emerging LPWAN technologies will help in building a reliable, low-cost, low-power, long-range, last-mile technology for smart energy metering solutions. We also present our prototype implementation of end-to-end LoRa connectivity for smart metering solution and discuss final visualization platform.

Keywords IoT · LPWAN · Smart metering solution · LoRa · DLMS protocol

---

S. Jain (✉) · M. Pradish
Central Power Research Institute, Bangalore, India
e-mail: jain.shobhit90@gmail.com

M. Pradish
e-mail: pradish@cpri.in

A. Paventhan
ERNET India, IIT Madras Research Park, Chennai, India
e-mail: paventhan@eis.ernet.in

M. Saravanan · A. Das
Ericsson Research, Chennai, India
e-mail: m.saravanan@ericsson.com

A. Das
e-mail: arindam.das@ericsson.com

© Springer Nature Singapore Pte Ltd. 2018
R. K. Pillai et al. (eds.), *ISGW 2017: Compendium of Technical Papers*, Lecture Notes in Electrical Engineering 487, https://doi.org/10.1007/978-981-10-8249-8_2

1 Introduction

Growing electricity demand, distributed generation, integration of renewables, advancements in IoT networks which attempts for managing the power system network with various levels of automation, managing outages, etc., have focused attention on the role of converting the electric grids to be smarter and efficient. One way, it helps in increasing electricity reliability—especially by increasing the grid system flexibility for various solutions. The smart grid, which overlays the traditional electrical grid with an IoT network that includes smart meters, will help in managing and monitoring various parameters of consumption pattern of customers on near real-time basis. A smart grid needs to introduce the transformation within the traditional electricity network with a series of new smart technologies. These include smart sensors, new backend IT systems, smart meters, and a communications network [1]. Moreover, it explores different technologies to automate the entire grid operation which includes power generation, transmission, and distribution [2]. For example, smart grids are used to improve the precision in operation and lowering the costs involved during the time of deployment of electricity infrastructure in developing countries. The introduction of small "remote" systems for rural electrification considered as a cost-effective approach can easily be extended to national or regional infrastructure [3]. But extending this to developing countries seems to be really challenging task due to huge infrastructure cost involvement for installation and recreating new environment. Some of the unlicensed technologies like LoRa, Sigfox and cellular technologies like LTE-M, NB-IoT are actively being deployed on a pilot basis in developing countries as a proof of concept without changing the complete environment. In utility sector, implementing such pilot projects will depend on the consumer requirements, geographical area, and the existing system based on the suitable technologies and standards that are adopted for achieving seamless end-to-end connectivity.

In this paper, our approach is to apply the emerging technology to the existing energy meter system that can communicate in real time or near real time benefiting both consumers and utility. So, we anticipate that the proposed system will enable the smart meter functionalities like automated meter reading (AMR) to utility companies using any last-mile connectivity on a near real-time basis. The challenges in such implementations in rural areas, is the requirement of the low-power and long-range connectivity with minimum installation cost. Hence we have chosen the LPWAN technology for the existing energy metering system.

LPWAN technologies allow utility sector to receive information over much longer range than traditional solutions in both rural and urban areas as we tested with our prototype for smart metering solutions. Moreover, in this paper we describe the metering architecture using LoRa for collecting data from the meter through low power microcontroller using RS-232 communication. LoRa supports two-way wireless communication for the metering infrasrucutre connectivity to the cloud through the gateway for realizing smart metering solutions.

Rest of the paper is organized as follows: Sect. 2 represents the background of technology and metering protocols, Sect. 3 describes LoRa components used in our prototypes, and Sect. 4 discusses the implementation setup of prototype demonstration and the results. The concluding remarks are presented in Sect. 5.

2 Background

With recent advancements in IoT networks, globally the power utility companies are deploying smart meters with various communication technologies. LPWAN technologies are the one that is extensively available in today scenario for a long range which can be leveraged for the smart metering implementations. Energy meter could be configured with this smart module and achieve most of the benefits of smart meters. The data transmission uses low-power, long-range, and narrowband transmission resulting in increased range of operations, better network stability, reliability, and optimum cost for implementation. This module could be configured to give personalized notifications at programmed intervals to the utility and the consumers. In this section, we describe the background details of the technology used and the relevant metering protocols.

LPWAN technology is used when other wireless networks are not a good fit with some case studies—Bluetooth and BLE (and, to a lesser extent, WiFi and Zigbee) and even that are often not suited for long-range performance. Cellular M2M networks are usually very costly, and it consumes lot of power as well as expensive as far as hardware and services are concerned [4]. The newly introduced LPWAN technologies are most suitable for this type of applications where the connecting devices generate small amounts of data and try to communicate over a long range, while maintaining long battery life [4]. We would like to discuss in this paper two main areas that best suits for LPWAN technology implementation relevant for smart grid automation:

1. Fixed, medium-to-high density connections which are alternative to the cellular M2M connections for introducing automation in grid distribution.
2. Long life, battery powered applications which are best fit for connecting the energy meters to substations and for other relevant purposes like water metering, gas detectors, smart agriculture, and operating door locks and access control points in home and office environments.

Sigfox [5] and LoRa [6] are the prime competitors in the LPWAN space for last two years. While the business models are quite different, the technologies and end goals are very similar. Both works for IoT application with low-power, long-range network with limited differences. Sigfox is a narrowband technology that uses a standard radio transmission method called binary phase-shift keying (BPSK), and it works with very narrow parts of spectrum and changes the phase of the carrier radio wave to encode the data [5]. It requires an inexpensive end-point radio and a more sophisticated base station to manage the network [7], whereas LoRaWAN [6] works at a wider amount of spectrum. NB-IoT is a low-power wide area (LPWA) technology that works virtually anywhere using cellular networks by generating more secure transactions [8]. It connects devices efficiently on already established mobile networks and handles low bandwidth two way communication, securely and reliably. This next-generation cellular IoT technology is not ready for the deployments till 2018 and needs operators support for end-to-end connectivity.

In the current scenario, 2G networks are not fully suited to IoT applications due to battery life constraints as well as higher complexity. NB-IoT and LTE-M technologies are not ready for immediate implementations due to the licensed spectrum issues and standards are still being developed. Whereas, LoRaWAN uses unlicensed spectrum by encoding multiple bits per symbol with integrated packetization and error corrections. The benefits of LPWAN technology with respect to LoRa can be presented here by mentioning the merits and facts [9].

(a) Spectrum: It uses unlicensed spectrum (865.20–867 MHz in India and 915 MHz in US).
(b) Long Range: With 2–4 km in dense urban areas and up to 15–30 km in rural areas. (depends upon line of sight).
(c) Power: 10+ years of battery life (The device's power produced cannot exceed 10–25 mW, to comply with usage of ISM frequencies, and to limit data consumption and preserve battery life).
(d) Radio chipset cost: $2 or less.
(e) Radio subscription cost: $1 per device/year.

Existing wireless networks are already established technologies, and they have well-established standards such as WiFi, Bluetooth 4.0, Zigbee, and Z-Wave [6]. Some of the issues with local area networks/personal area network are the battery consumption and connectivity range. Similarly, the cellular networks such GSM, 3G, 4G, LTE are also established proprietary mobile networks developed for better network coverage and data throughput, but it is not considered best when it comes to power consumption. On the other hand, LoRaWAN is a platform that can be built according to the specification. LoRa is exclusively developed to work with IoT devices which needs better battery life, low data transmission and minimal cost for deployment. On other hand, LAN and cellular network are quite expensive to deploy in a wide area, whereas LoRa usage would be much easier and cost effective for the implementation, and it follows the open standards. Next, we will discuss the relevant metering technology which supports our implementation.

Device Language Message Specification (DLMS) is an application layer protocol supporting multiple transport layer options including Ethernet and PLC. Companion specification for Energy Metering (COSEM) specifies the data model with their attributes and methods [10, 11]. DLMS/COSEM is open standard managed by DLMS user association. This protocol is accepted for use in most of the metering arena across the world. Moreover, it is accepted by most energy utilities in the wake of liberalized energy market. Several countries including India have framed metering specification based on DLMS which will act as a guideline for utilities to specify their metering requirements and ultimately improve consumer data collection, billing, and revenue.

OBIS code is the vital component used within DLMS protocol supported meters [12]. OBIS, an Object Identification System, provides the standard identifiers for all data within the metering equipment, both measurement values and abstract values. Moreover, the OBIS names are used for the identification of COSEM objects and the data displayed on the meter and transmitted through the communication line to the data collection system [12]. Easily, we can interpret the values extracted from meters through OBIS codes. The extracted values are communicated to the cloud through LoRa module used in our implementation. Now, using this new technology, we can extract the multiple relevant parameters from the general energy meter used in households for the analysis. We have also designed an analytical model for recommending certain schemes relevant to the consumers in a cloud setup.

3 LoRa Module Components

In our prototype, in order to connect the energy meter for reading and communicating, we have employed LoRa module which consists of two main interfacing components: a microcontroller that is attached with a LoRa RF module to get the details out of the meter and LoRa gateway for aggregation as shown in Fig. 1. The microcontroller block can be considered as the main block of the entire circuit, as it is programmed to control all the components to perform the desired operation. Here, in this project ARM® Cortex® microcontroller along with Semtech SX1272 LoRa extension board and programmed using the Arm Mbed OS platform. The energy meter is interfaced with the microcontroller using a serial converter device compatible with the meter's serial port. Similarly, LoRa modem is interfaced with a microcontroller to GPIO pins using interfacing connectors and cables.

LoRaWAN architecture uses a star topology network in which gateways acts as a transparent bridge relaying messages between end-devices and connects to the central network server in the backend [8]. In this setup, gateways are connected to the network server through a standard IP connection and the end-devices use single-hop wireless communication to one or many gateways [7]. Here, all end-point communications are generally forced to be bidirectional and it supports operation such as multicast enabling software upgrade over the air or other mass distribution messages to reduce the on-air communication time [8]. The communication between the end-devices and gateways are based on different frequency channels and different data rates [13, 14].

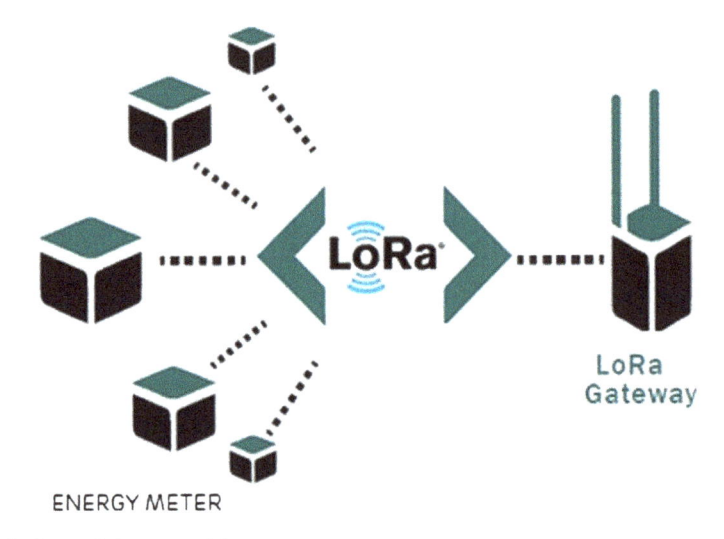

Fig. 1 LoRa module connectivity to gateway

4 Implementation and Discussion

For the experiment, we have taken a DLMS-based energy meter to interface the same with a microcontroller-based module that can transmit the meter parameters to our controller and transmit the same through LoRa RF module. The base module is interfaced with the energy meter through interfacing modules and serial port connections. LoRa module can read the serial data from the meter through serial communication device, and the same is then transferred into machine-readable values using OBIS codes. The initial setup of the meter involves configuring baud rate (9600 bps) and Rx, Tx pins for subsequent communications. The data received from the microcontroller through Rx, Tx ports is transferred using the LoRa communication interface as shown in Fig. 2.

We have shown here two code snippets, one to demonstrate the connection of serial communication with microcontroller from meter (Fig. 3) and the other to show the relevant data fetching from DLMS meter using OBIS reference code (Fig. 4).

The microcontroller needs to first establish the connections with the meter through sending OBIS request for the first time. Once the energy meter is setup in the environment, the required OBIS code is sent to get various parameters of data.

The values received from the meter are not in machine-readable format, as those data are encoded as specified in DLMS protocols. The encoded data is then converted using OBIS code reference model into readable format. The data is then transferred to LoRa gateway for the further processing as shown in Fig. 5.

The data received from LoRa module is thereafter transferred to LoRa gateway for uploading the same in the cloud. The distance between gateway and the modules

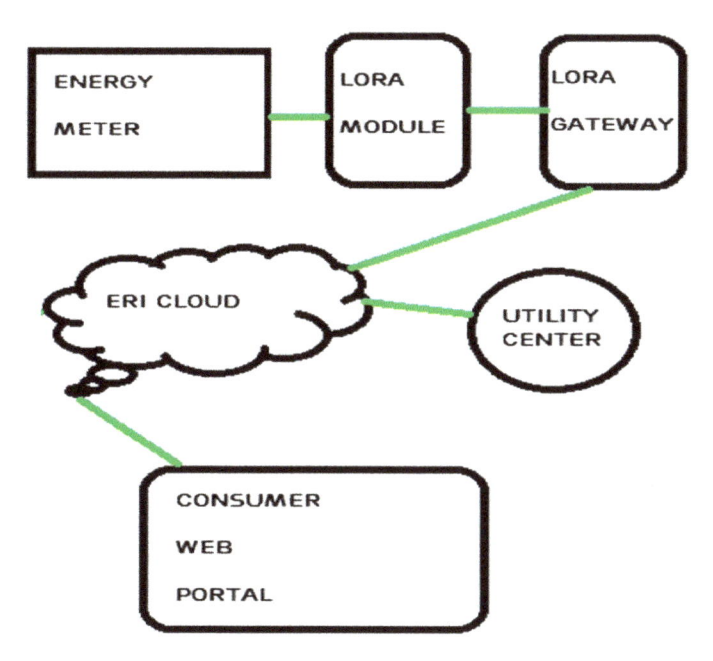

Smart Energy Metering Using LPWAN IoT Technology 25

Fig. 2 Energy meter connect to LoRa module

```
/*-----( Import needed libraries )-----*/
#include <SoftwareSerial.h>
#include <ctype.h>
/*-----( Declare Constants and Pin Numbers )-----*/
#define SSerialTxControl 3 //RS485 modual control pin
#define RS232Transmit  HIGH
#define RS232Receive   LOW
#define Pin13LED 13

  /*-----( Declare objects )-----*/
SoftwareSerial RS232Serial(10, 11); //10 is RX and 11 is
TX

  /*----- (Declare Variables) -----*/
String readOut[20] = "";
int BaudRateL[4] = {1200, 2400, 4800, 9600};
char BaudRateM[4] = {0x32, 0x33, 0x34, 0x35};
//Hex value for sending your preferred baud rate to the
Meter.
```

Fig. 3 Code snippet—RS232 communication setup

```
// ****** SETUP: RUNS ONCE ****** //
Void setup () {

//Need to set the number of data points we are
going to use and set them all to 0
//ensures i start with blank data which i can omit
later.
//These are strings is i got results like "0.004*Kw"
from the meter.
  int z = 0;
  while (z < 6) { // if you send 10 OBIS codes then
set 6 to 10, etc.
    readOut[z] = "0";
    z++;
  }
}
```

Fig. 4 Code snippet—data pooling setup

is configured in such a way that a gateway can handle more than 10 LoRa module in an area of 4–5 kms, and each LoRa module can be attached with at least 10 m.

The data received on gateway is then transferred to the Cloud setup, for putting those into prediction models. The data from the cloud is then transferred to two places—firstly into the utility centers with a provision of sending commands back to the microcontroller and another to the Web page for displaying to customers. The newly designed Web page for the smart metering solution is shown in Fig. 6.

Ericsson Cloud (Eri Cloud) [15] enables to transform the data back to LoRa module for actuation based on analytics. It also stores the data collected from the gateways, that is accessible and available in real-time.

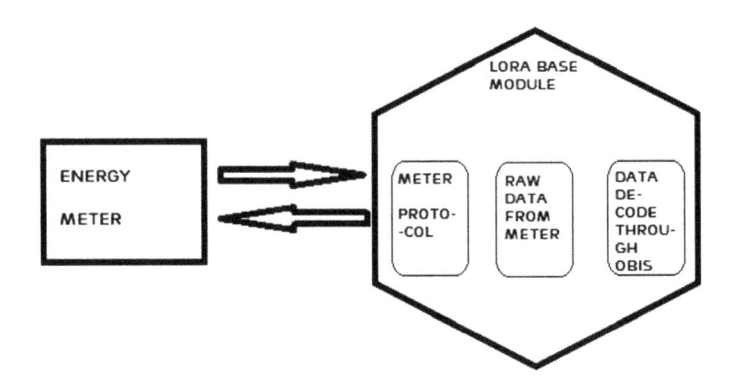

Fig. 5 Data conversation method

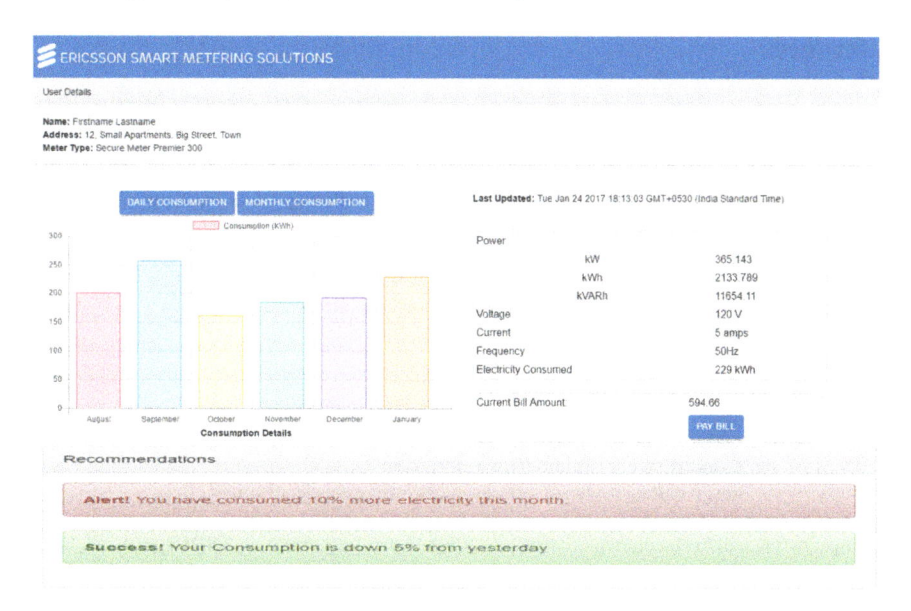

Fig. 6 Web portal for consumption details

The Web page is connected with the Cloud to get the real-time updates out of the LoRa gateway and display the same. It also shows the results of prediction models that run on stored data. These results help the consumer to track down their usages and also to have limited control on the devices. The administrative part of the Web portal is controlled by the utility centers.

5 Conclusion

In this paper, we have demonstrated the LoRa implementation for the DLMS-based energy meter for establishing communication and data extraction. Our prototype implementation showed how LoRa as a last mile technology can benefit for the smart metering solutions. The metering data can be moved to the cloud for storing and running prediction model to generate consumer-specific recommendations. The energy meter parameters like voltage, current frequency, power factor, and consumption can be directly accessed through GUI, and also it can be directed to consumer mobile through mobile app for decision making and controlling the device usage. Our prototype can easily be extended in the real-time environment for the benefit of customers and utility centers to provide smart metering solutions.

References

1. https://www.cookinghacks.com/documentation/tutorials/how-to-send-sensor-data-using-lora-extreme-range/
2. Smart Meters and Smart Meter Systems, A Metering Industry Perspective. An EEI-AEIC-UTC White Paper; February 2011 IEC 62056, "Electricity metering—data exchange for meter reading, tariff and load control," International Electrotechnical Commission series of standards, 2002
3. Sornin N, Luis M, Eirich T, Kramp T, Hersent O (2015) LoRaWAN specification version 1.0
4. http://cdn2.hubspot.net/hubfs/427771/LPWAN-Brochure-Interactive.pdf
5. http://www.link-labs.com/blog/sigfox-vs-lora
6. http://www.3glteinfo.com/lora/lora-advantages/
7. www.newark.com/wireless-sigfox-technology
8. https://www.lora-alliance.org/For-Developers/LoRaWANDevelopers
9. https://www.semtech.com/technology/lora/what-is-lora
10. DLMS User Association (2013) http://www.dlms.com
11. https://www.dlms.com/faqanswers/questionsonthedlmscosemspecification/obisnameswhatarethey.php
12. https://www.kalkitech.com/wp-content/files/WhitePaper_Implementing_DLMS_protocol_in_meters.pdf
13. http://www.3glteinfo.com/lora/lorawan-frequency-bands/
14. http://www.postscapes.com/long-range-wireless-iot-protocol-lora/
15. https://www.ericsson.com/hyperscale/cloud-infrastructure

Smart Grid Technologies: Distribution Automation, Microgrids, and Cyber Security

S. R. Vijayan

Abstract The recognition of the contributions and challenges of the distribution system for delivering the generated power to the end consumer with high availability, reliability, and efficiency has increased the responsibility of the distribution system owners and operators. Accordingly, this segment of the entire delivery chain is now seeing increased focus on developing and strengthening the network both in terms of equipment installation and implementing automation solutions.Further as the percentage or penetration of renewables increases in a power system, it becomes imperative to have grid level power balancing using energy storage systems. Microgrids are expected to form an inherent component in the grid system with localized generation and storage as close to the load centers as possible. It also makes sense to encourage shifting the loads to a time of the day when there is surplus energy available in the system through "Demand Response." The urgent need of making the grid "SMART" has made the Operational Technology (OT) systems which were otherwise in a secure network to handshake with the external systems for data and information exchange. This induces a sense of insecurity to the operational assets of the utilities. Thus, when a utility is implementing various solutions and integrating these systems for data exchanges, the systems need not only be interoperable but also need to be secure. This technical paper will discuss the various Distribution Automation technologies, the Microgrids concept, and the cyber security vulnerabilities and mitigation techniques.

Keywords SCADA/DMS · Microgrids · Outage management system (OMS)
Advanced metering infrastructure (AMI) · Cyber security · Demand Response

S. R. Vijayan (✉)
Power Grids—Grid Automation, ABB India Ltd., Bangalore, India
e-mail: Vijayan.sr@in.abb.com

© Springer Nature Singapore Pte Ltd. 2018
R. K. Pillai et al. (eds.), *ISGW 2017: Compendium of Technical Papers*, Lecture Notes in Electrical Engineering 487, https://doi.org/10.1007/978-981-10-8249-8_3

1 Introduction

The modern-day power system is transforming over the recent years from the traditional unidirectional (generation to distribution to consumers) power flow to an interconnected grid, with distributed generation injecting power into the grid at different levels (sometimes even at the consumer point) within the system (Fig. 1).

The distribution utilities today face challenges not only from the grid connectivity point of view but also in delivering power with increased efficiency and reliability. With the distributed generation, the grid connectivity and the power flow are changing and changing fast from a radial, unidirectional flow network to an interconnected and multidirectional flow network. Variability of solar and wind power in a small geographical domain calls for quickly controllable backup generation or storage such as battery systems to fill in shortfalls.

Under these circumstances of operating a distribution system, automation solutions become an effective tool not only for the operators to monitor and control the network but also for the management to take decisions. The grid automation provides a greater and wider visibility of the network, thereby enabling remote and faster controllability. Implementation of automation brings in enormous data from the field, and there is a growing importance of managing these data and using them as information by implementing Information Technology (IT) solutions.

As the percentage or penetration of renewables increase in a power system, it becomes imperative to have grid level power balancing using energy storage systems. Microgrids are expected to form an inherent component in the grid system with localized generation and storage as close to the load centers as possible.

Traditional grid – Relatively Simple

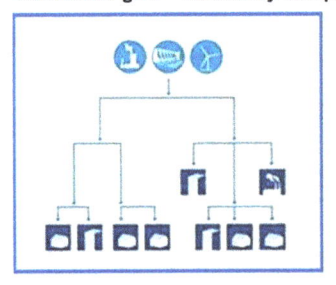

- Centralized power generation
- One-directional power flow
- Generation follows load
- Top-down operations planning
- Operation based on historical experience

Interconnected grid – New Complexities

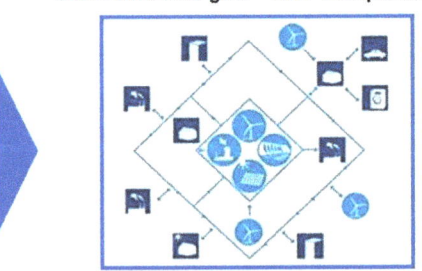

- Centralized and distributed power generation
- Intermittent renewable power generation
- Multi-directional power flow
- Consumption integrated in system operation
- Operation based on real-time data

Fig. 1 Traditional versus interconnected grid

The traditional control systems which are called the Operational Technology (OT) need to be bridged with the IT systems. While a smart grid converges the OT and IT systems within an utility, the important aspect of its implementation is to efficiently integrate them as per the utility requirements.

As the IT and OT systems are getting integrated, the OT systems which were otherwise in a secure perimeter (physical and electronic) to handshake with the external world, thereby inducing a sense of insecurity to the operational assets of the utilities. Also using the Ethernet-based technology in the automation systems is gaining wide popularity and acceptability, as it gives tremendous benefits in terms of solution implementation and maintenance. They also bring in the threat of viruses, Internet-based attacks, etc. Thus, when an utility is implementing various solutions and integrating these systems for data exchanges, the systems need not only be interoperable but also need to be secure. Hence, the cyber security for automation in electric utilities has gained a lot of importance and attention in the recent years. Cyber security has transformed from a nice-to- have to a must- have for the utility world over.

2 Distribution Automation

While distributed generation is changing the traditional grid connectivity, the distribution system needs implementation of technologies to improve its operational efficiencies like energy availability, power quality, and system responsiveness. To achieve this, a smart utility has to not only monitor the system components at every location of the distribution system but also plan "automatic" reconnection of disconnected customers, when technically feasible, within a short time. The automation solutions include applications like SCADA/DMS, outage management system (OMS), advanced metering infrastructure (AMI), smart metering, and advanced applications like Demand Response.

While SCADA is the basic platform of an automation system, the applications for the distribution network widely known as Distribution Management System (DMS) are a key component of smart grid (or) Distribution Automation. The DMS provides functionalities to improve the operations and the efficiency of the sub-transmission, medium- and low-voltage distribution networks. The DMS interfaces with the SCADA system and facilitates the network analysis and network optimization of the distribution system. Typical system architecture of a Distribution Automation system with the IT-OT convergence is shown in Fig. 2.

There are different traits that a Distribution Automation is expected to improve in the power system operations as part of the smart grid solutions, like:

- Optimize asset utilization and operating efficiency.
- Provide power quality.
- Anticipate and respond to system disturbances in a self-healing manner.
- Operate resiliently against physical and cyber attacks.

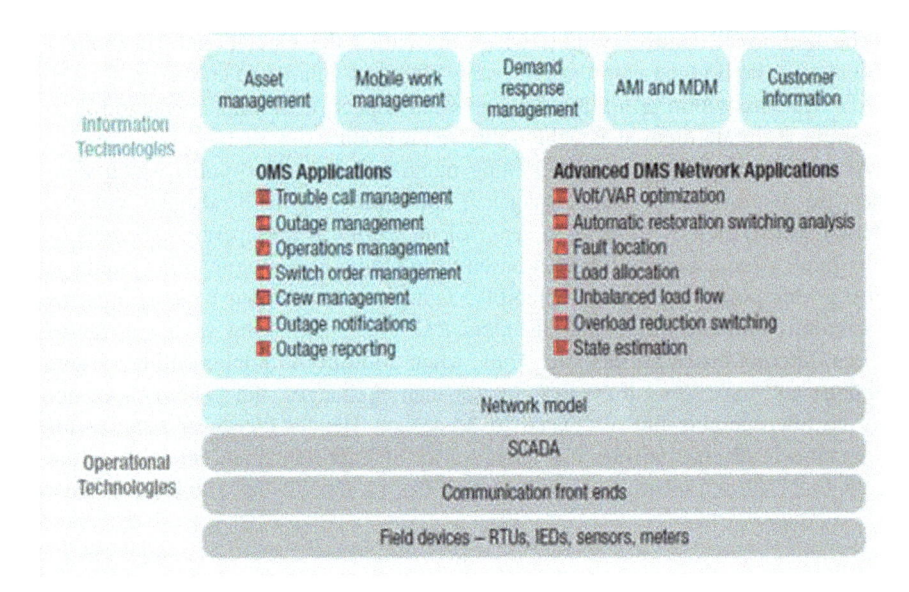

Fig. 2 IT-OT convergence in Distribution Automation

- Enable active participation by consumers.
- Address distributed generation and energy storage options.

The reliability of a distribution system is measured by two main indices—customer average interruption duration index (CAIDI) and system average interruption duration index (SAIDI). The efficiency of a distribution system is measured by the distribution losses and the voltage quality.

Accordingly, the two main applications of the DMS are

- Fault location, isolation, and restoration (FLISR)
- Volt–var optimization (VVO)

In FLISR, the fault location deals with detecting the permanent fault in a feeder. The faults are detected through the relays whose values are telemetered through the SCADA system. The fault parameters are read and analyzed by the DMS applications and based on the network topology built in the system at the time of system engineering, faulty sections are identified. Once, the fault and the faulty section are identified/located, the isolation function isolates the faults by producing operation sequences (also called as switching sequences) of the network devices which can be remotely controlled. The restoration function attempts to restore the power to as many customers as possible, thereby reducing the number of customers affected and improving the reliability indices. The restoration function also produces switching sequences of the network devices which can be remotely controlled for supply restoration. These switching sequences can be executed either automatically or manually. In automatic switching, the switching operations are executed automatically with/without operator approval, depending on the system configuration.

In case of manual switching, each operation in the switching sequence for isolation/restoration has to be done by the operator. The automatic/manual operations depend on the utilities' needs and possibilities.

The volt–var optimization (VVO) function improves the efficient operation of the network by:

- Keeping voltages in limits
- Reducing losses by minimizing reactive power flows
- Reducing peak power by voltage reduction.

The capacitor banks and the transformer tap changers are controlled to achieve the required voltage and reactive power operating range. This application is also interfaced with the SCADA system and uses the real-time network information, thereby using the "as-is" status of the network in the optimization calculations.

2.1 Outage Management System (OMS)

While a SCADA/DMS system improves the overall operational reliability and efficiency of the network, there is always possibility of outages. These outages can be either planned or unplanned. The OMS system provides a platform to intelligently manage the outages and provides support in decision-making process to support daily grid operations and minimize outages.

2.2 Advanced Metering Infrastructure (AMI)

AMI is a complete infrastructure which includes smart meters, communications, meter data management system (MDMS). In the implementation of AMI, the communication is bidirectional, which means that this facility enables monitoring the consumption of the customer connected as well as controlling of the meter/appliances. In fact, the bidirectional communication is the major difference between the traditional automated meter reading (AMR) system and the AMI. The smart meter component of an AMI system also has a display unit either within the meter or separately mountable unit, which displays information about the energy usage. With this information, customer can make choices of running the home appliances selectively and efficiently based on the consumption pattern and the price at the time of use. The communication network which binds the distribution system with the customer and "inside" customer premises is called home area network (HAN) and neighborhood area network (NAN).

AMI, while enabling bidirectional communication between the utility and the customer, can also be extended to a full fledge Demand Response program.

2.3 Communication Technologies for Distribution Automation

Communications between the field devices and the control center executing the automation applications have a critical role in the successful implementation of automation systems. Further, in efforts to reduce the fault restoration time in FLISR, horizontal communication or the peer-to-peer communication is becoming very much of an implementation philosophy of smart grid solutions. These are popularly known as self-healing grids. The importance of evaluating the system's reliability, security, and availability as well as technology cost when choosing communication technologies cannot be underrated. The applications like FLISR used for reducing the outages and affected consumers by remote operation of ring main units (RMUs) and the volt–var optimization are predominantly used applications in the Distribution Management System.

In terms of expanding the automation coverage of the distribution network, the enablement of the field devices to communicate with the control center is a key factor. Further, the expansion can also be in terms of application implementation itself like AMI being done phase-wise. Hence, the communication system should be strategically planned to allow scalability of the system. Besides the AMI, even the RMU automation can be taken up in a phased manner to increase the network visibility at the control center. The ability to scale up the communication network becomes an important parameter.

From the communication perspective, the entire chain between the control center, the substations, RMUs, the smart meters, and the home area Network (HAN) the communication layer can be tired into four basic layers to cater to the Distribution Automation needs. Figure 3 indicates these four layers.

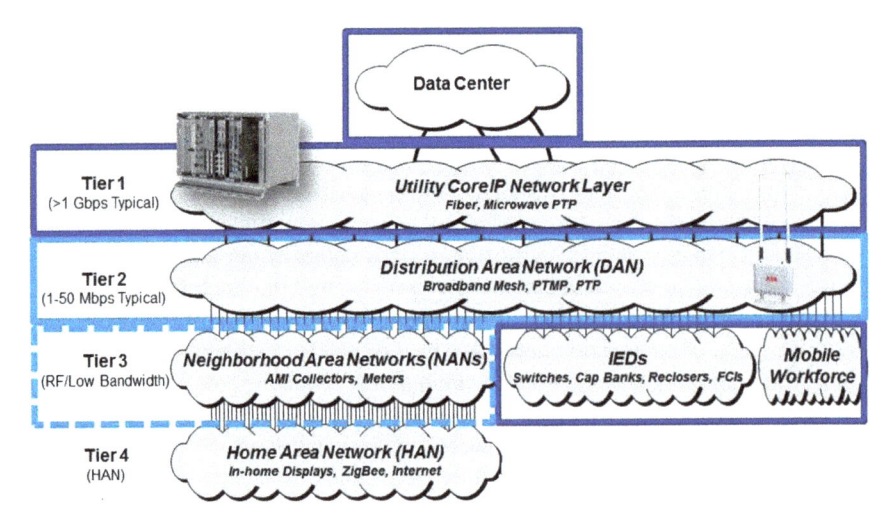

Fig. 3 Communication tiers in a DA system

Smart Grid Technologies: Distribution Automation ...

As Tier 3 and Tier 4 become more distributed as the geographical reach is spread across, providing a robust Tier 1 and Tier 2 network becomes a key factor in highly enhancing the reliability and availability of the DA system. While it is comparatively easier to choose a fiber optic or a microwave radio network for the Tier 1, it is difficult to choose the right technology for Tier 2.

A comparison between the various communication technologies for the Tier 2 is given below:

Technology capability	Landline	Wireless telecom	Power line carrier	Private radio mesh
Infrastructure available in India	Remote area connectivity not available everywhere	Depends on the service provider in a particular location	Dependent on the reliable distribution network. Not widely used in India until now	Mesh networks can be established based on the needs
Cost of transferring data	Low	Low (based on transmission during off-peak time)	Low (different standards must be implemented first so initial cost may be high)	Free
Reliability	Low (due to risk of physical damage)	High, however, depends on the service provider in a particular location	Low (due to use of newer and untested technologies)	High
Risk involved	Physical damage	High traffic during daytime	Newer, relatively untested technology	Onsite support
Scalability	High (major investment needed)	Low (bandwidth in wireless domain is limited)	High (major investment needed)	High (technology is not a barrier)

Application bundling in the existing network	Not possible	Not possible (bandwidth in wireless domain is limited)	Not possible	Very High
Cyber security	Secured	Public network—VPN required	Secured	Secured

From the above comparison establishing a private radio mesh network seems to be a strategic communication option looking on the long-term benefits it stands to deliver.

3 Microgrids

The variability of solar and wind power in a small geographical domain calls for quickly controllable backup generation or storage such as battery systems to fill in shortfalls. In urban scenario, the loads may peak during morning and evening hours when there is no solar generation.

As the percentage or penetration of renewables increase in a power system, it becomes imperative to have grid level power balancing using energy storage systems. Microgrids are expected to form an inherent component in the grid system with localized generation and storage as close to the load centers as possible. They can be differentiated from major power grids in the sense that they are usually integrated at the distribution system of the power system, limited to integration at low- or medium-voltage levels of the grid. They are limited in geographical reach unlike power grids, and they lack bulk power transmission capabilities. Microgrids are thus more localized where the generation and consumption happen within a small area.

One may visualize Microgrid as a part of a power system with generation, having a mix of renewable and conventional, with storage elements for grid stabilization added (if a high penetration of renewable has to be achieved within the Microgrid so formed) enabling independent operation to be achieved with or without provision to be connected to a grid as indicated in Fig. 4.

Microgrids being smart themselves can operate independently in remote communities but when multiples of them get integrated with the powergrids, they form the basic building blocks of a smart grid at distribution level forming smart grids. They are limited in geographical reach unlike power grids, and they lack bulk power transmission capabilities. Microgrids are thus more localized where the generation and consumption happen within a small area.

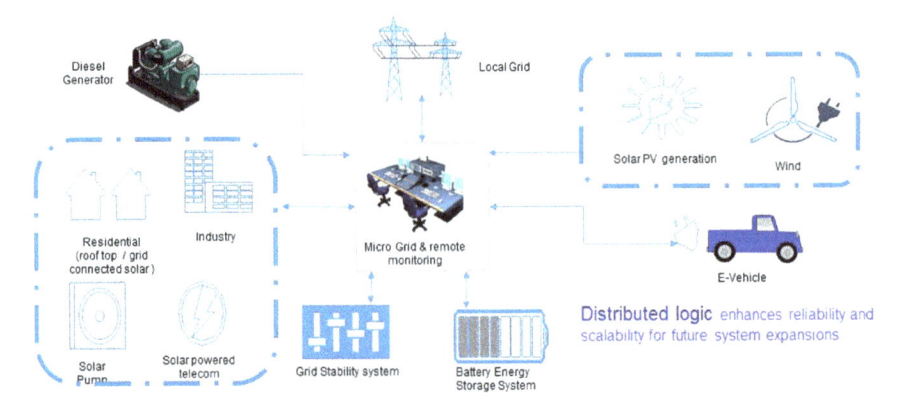

Fig. 4 Interconnection of microgrid components

			Main drivers				
			Social	Economic	Environmental	Operational	
Segments	Typical customers		Access to electricity	Fuel & cost savings	Reduce CO2 footprint and pollution	Fuel Independence	Uninterrupted supply
Island utilities	(Local) utility, IPP*			✓	✓	✓	(✓)
Remote communities	(Local) utility, IPP, Governmental development institution, development bank		✓	✓		✓	
Industrial and commercial	Mining company, IPP, Oil & Gas company, Datacenter, Hotels & resorts, Food & Beverage			✓	(✓)	✓	✓
Defense	Governmental defense institution			(✓)	(✓)	✓	✓
Urban communities	(Local) utility, IPP				(✓)		✓
Institutions and campuses	Private education institution, IPP, Government education institution			(✓)	✓		(✓)

IPP: Independent Power Producer
✓: Main driver
(✓): Secondary driver
Source: ABB

Fig. 5 Some main driving forces in applying Microgrids

The motivations for installing and operating a Microgrid vary depending on the segment in which the power system is applied. Figure 5 shows an overview of when and where a Microgrid could be considered depending on various factors.

3.1 Demand Response

Demand Response is a program to control and change the end customer energy consumption pattern depending on the factors like time of use price, peak shave-off incentives. Peak shave-off is reducing the consumption to reduce and bring down the peak levels, so that the electrical network can deliver without any disturbances arising due to overloads. As peak shave-off is more important from the utility perspective to maintain the operating conditions of the grid, incentive payments are normally included to encourage end customers to volunteer and subscribe to the program.

Figure 6 depicts the demand curve with and without the Demand Response implementation. The time-of-use consumption will trigger self-interest in the end customer to use the appliances based on the tariffs fixed at different time of the day. DR opens up a disciplined usage of energy by intentionally altering the time of usage of home appliances. Demand Response is a voluntary program where the end customer "participates" in helping the utilities to manage the energy distribution.

The combined solutions of Microgrids and AMI systems enable effective implementation of Demand Response. With the AMI implementation encompassing residential consumers as well, the Demand Response program with smart meters will act as a motivational factor for them.

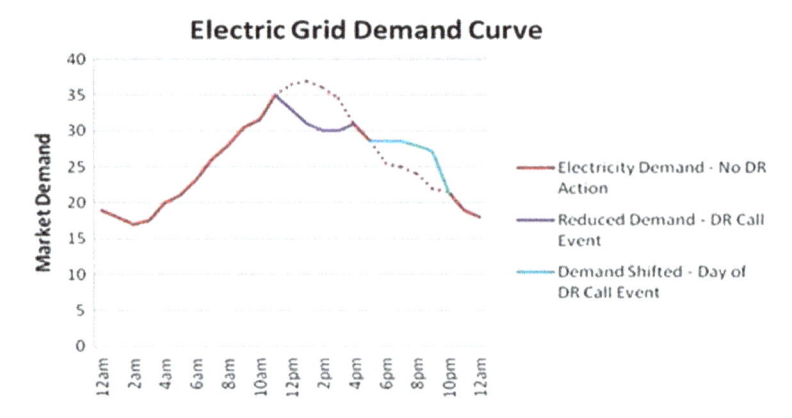

Fig. 6 Demand curve behavior and Demand Response

4 Cyber Security

As we move forward with implementing intelligent solutions and making the grid smart to deliver quality power with improved efficiency, reliability, and in a sustainable way, there is also a major concern. Building intelligence means interconnecting various devices and systems to exchange information to bring in more situational awareness about the status of the network and decision-making support to operate and maintain the grid. With implementing solutions like Demand Response, the devices getting connected are no more limited to the utility premises but include devices and appliances in the customer premises.

As the IT and OT systems are getting integrated, the OT systems were otherwise in a secure perimeter (physical and electronic) to handshake with the external world, thereby inducing a sense of insecurity to the operational assets of the utilities. Also using the Ethernet-based technology in the automation systems is gaining wide popularity and acceptability, as it gives tremendous benefits in terms of solution implementation and maintenance. They also bring in the threat of viruses, Internet-based attacks, etc. Thus, when a utility is implementing various solutions and integrating these systems for data exchanges, the systems need not only be interoperable but also need to be secure. Hence, the cyber security for automation in electric utilities has gained a lot of importance and attention in the recent years. Cyber security has transformed from a nice-to- have to a must- have for the utility world over.

Recognizing the need of building the security into the deployments, the utilities have started implementing solutions and execute operational practices toward addressing the cyber security aspect. It will be worthwhile to note that, addressing the cyber security is not limited to building different technical solutions but it is also as much important to have strong operational (or) IT policies within the utility organization.

There are various methods to prevent a cyber vulnerability in a control system. Some of these methods are mentioned below:

System Hardening protects the field devices as well as the control center devices by blocking the unwanted software, services, and ports.

Role-Based Access Control (RBAC) extends the basic user authentication mechanism to allow access of the authorized users to the level restricted based on their roles.

Network Partitioning is the method to define the electronic perimeter of the various network segments implemented based on the needs. The defined perimeters are then fenced using firewalls. This includes creating a de-militarized zone (DMZ) for access by external users.

Intrusion Detection and Prevention Systems enable monitoring of the network traffic and connected devices and create security logs and alerts to respond to possible anomalies in the system.

Patch Management mechanisms help the user in maintaining the system and keep them updated with the latest security updates that are verified. The patch management includes security patches for anti-virus, operating systems, applications, tools, etc., as may be applicable for a specific installation.

Anti-virus is a well-known and basic measure for protecting the systems from virus and malware attacks.

Application and Traffic whitelisting are the system definition to permit only pre-approved applications, software, and network traffic within a system.

While the need and the importance of this critical aspect of automation are felt across the utilities, the drivers and the level of implementation may differ. However, the demand for cyber security solutions will increase and become mandatory requirements as part of solutions and products. Accordingly, the standards organizations are also taking up the standard's development activities on priority. The table below summarizes the various standards around the cyber security.

Standard	Main focus
NIST SGIP-CSWG	Smart grid interoperability panel—cyber security working group
NERC CIP	NERC CIP cyber security regulation for North American power utilities
IEC 62351	Data and communications security
IEEE PSRC/H13 & SUB/C10	Cyber security requirements for substation automation, protection, and control systems
IEEE 1686	IEEE standard for substation intelligent electronic devices (IEDs) cyber security capabilities
ISA S99	Industrial automation and control system security

In spite of having different methods to protect the systems against the cyber threats and vulnerabilities, implementing solutions to address cyber security needs to be a continuous one. The system protection can only be enhanced from one level to the next and the utility organizations cannot be complacent w.r.t cyber security.

Also, it is not sufficient to implement technical solutions to address this issue, where the organizational policies also need to be in place and reviewed time-to-time to have an effective protection of the utility assets.

5 Conclusions

The smart grid is more than any one technology, and the benefits of making it a reality extend far beyond the power system itself. While the transition in the grid connectivity is already happening and the utilities are facing the reality, the transition in the way this new gird is operated is the need of the hour.

However, this transition will not be easy. The integration of smart technologies of many different kinds will be essential to a functioning of a smart grid, and the path to integration is supported with interoperability standards.

The basic concept of Microgrid and its application in evolving Indian power scenario is highlighted. Microgrids are expected to form an important component of smart grid. They improve the operational efficiency as well as the reliability of the overall power system even during emergency scenario, enabling maintenance of lifeline of power and communications with great portions of affected systems.

Demand Response program provides a technical solution to the saying "Energy saved is energy generated." While it provides the power system with the required energy to meet the demand without a feeder level load shedding, it also a leaves a satisfied customer due to the incentives and the availability of power to meet his critical requirements. Thus, the implementation of DR is mutually beneficial to the utilities and the consumers.

Communications are the backbone of the successful implementation of Distribution Automation solutions and choosing the right one for today's and future requirements is the key for an efficient and cost-effective implementation. With choices of technology available, choosing the right one has to consider the long-term plans in implementing solutions in phases and weigh the pros and cons of each technology. While the popularly used GPRS network is easy for implementation and involves less initial costs, a wireless mesh indicates a highly reliable, scalable, and secured network for real-time data transfer for the Tier 2 network.

From the cyber security perspective, technology is insufficient on its own to provide robust protection. Cyber security policies and processes must be implemented in the organization to best be able to assess and mitigate the risks and respond to incidents. There is no such thing as 100% security or a 100% secured system. Hence, implementing solutions around the cyber security has to be a continuous one. Therefore, it is not only important to protect a system from the current vulnerabilities but also equally important to have mechanisms (technical and process) in place to quickly detect and effectively react to any incidents and isolate security breaches.

Impact Analysis of Attacks Using Agent-Based SCADA Testbed

M. Siddharth Rao, Rajesh Kalluri, R. K. Senthil Kumar, G. L. Ganga Prasad and B. S. Bindhumadhava

Abstract Supervisory Control and Data Acquisition (SCADA) systems are used to control and monitor the critical infrastructure such as electricity, gas, water. SCADA system networks are originally started as local control systems and have expanded to wide-area control systems. The integration of different networks leads to various cyber security vulnerabilities. Many of the SCADA systems are relatively insecure with chronic and pervasive vulnerabilities. Ever-growing threat of cyber terrorism and vulnerability of the SCADA systems is the most common subject for security researchers. With increase in both internal and external threats, it is required to analyze the impact of these attacks on SCADA system in terms of availability and performance. A testbed is needed as it is impractical to conduct any security experiments on a real SCADA system. This paper presents the experimental SCADA testbed using multi-agent framework. Simulation of attacks such as denial of service, man-in-the-middle attacks, and malwares can be conducted on testbed to analyze the impact of these attacks. Experiments have been conducted on

---

M. Siddharth Rao (✉) · R. Kalluri · R. K. Senthil Kumar ·
G. L. Ganga Prasad · B. S. Bindhumadhava
Center for Development of Advanced Computing (C-DAC), Knowledge Park,
No 1, Old Madras Road, Byappanahalli, Bangalore, India
e-mail: siddharthr@cdac.in

R. Kalluri
e-mail: rajeshk@cdac.in

R. K. Senthil Kumar
e-mail: senthil@cdac.in

G. L. Ganga Prasad
e-mail: gpr@cdac.in

B. S. Bindhumadhava
e-mail: bindhu@cdac.in

SCADA testbed by targeting performance and availability of the system, and the results can be studied using SCADA threat analyzer (STA) and security information and event management (SIEM) tool.

Keywords SCADA · MAF · CMAF · ReAS · RAS · PAS · STA
SIEM

1 Introduction

Supervisory Control and Data Acquisition (SCADA) systems are used to control large-scale processes that can be dispersed over long distances using centralized data acquisition and supervisory control. In the past, SCADA networks were completely isolated and used proprietary control protocols running on specialized hardware and software. The use of new IT systems provides improved connectivity across the SCADA network and remote access. By allowing the acquisition of data and control from remote locations, SCADA networks provide great efficiency. This in turn makes SCADA networks vulnerable to interruption of service, process diversion, or manipulation of data that could affect public safety or even severe disruption to a nation's critical infrastructure [1]. Furthermore, the goals of safety and efficiency sometimes conflict with security in the design and operation of SCADA systems.

To protect SCADA systems, it is essential to investigate the security threat of these systems and develop appropriate security solutions to guard them from attacks. A major problem in the development of security solutions for the SCADA system is the lack of appropriate modeling tools to assess the security of the SCADA system. A SCADA testbed provides a means to realize a SCADA system and also enables the testing of various security solutions on the SCADA system [2]. This paper defines the use of technology through multi-agent framework and targeting specific attack scenarios as an approach for analyzing attacks over SCADA tested. Further the analysis of malware attack over SCADA network is modeled by using a testbed with malware characteristics, security information extraction through a system-wide threat analyzer tool, and a single window alert system.

Section 2 delves into the SCADA testbed setup including agent-based system for simulation of attacks, analysis, and architecture of the SCADA testbed. Section 3 discusses the simulation of various attacks performed over the SCADA testbed and experimental results. Section 4 examines case study for impact of malware attack over a SCADA testbed by using the STA and SIEM tools.

2 SCADA Testbed Setup

A testbed is essential due to the fact that it is impractical to conduct security experiments on a real SCADA system because of the extent and price of implementing standalone SCADA systems, as well as the impending risk and downtime of services provided by the critical infrastructure controlled by the SCADA system. C-DAC multi-agent framework (CMAF) is used for simulation of the attacks and inter- or intra-system communication tasks over testbed. SCADA testbed is modeled to follow defense-in-depth architecture for layered security. The testbed is equipped with SCADA threat analyzer and security information and event management (SIEM) tool for analysis. The following sections describe these features in detail.

2.1 Agent-Based Environment for Simulation of Attacks

An agent is like a daemon acting on behalf of a user and performing some specific task in coordination with other agents. A mobile agent is a program, which moves around from node to node in a network. Mobile agents migrate autonomously (on their own) or on request from the external world from node to node in the network, to perform some task on behalf of the user. They can perform a wide variety of tasks, which can range from online shopping to real-time data device control to distributed scientific computing.

C-DAC's multi-agent framework (CMAF) has been adapted for simulation of attacks. C-DAC multi-agent framework is a network-based application developed in Java and is based on the concept of code movement across the network with or without data. It provides an agent system in which agent runs. It helps the software agents to execute, to handle their execution, to use system resources, and to assure reliability and protection of agents and the platform itself. It also provides support for migration, naming, location, and communication services as shown in Fig. 1. Following features are provided by CMAF: platform for agents, APIs to create agents, collaborative environment (agent-to-agent communication), pluggable services: communication, console/GUI, registry, adheres to MASIF and FIPA standards, agent mobility-hybrid (function-level), encryption: DES encryption, Self-CHOP (configuration, healing, optimization, protection), agent and agent system-level fault tolerance. CMAF has support for hybrid mobility, a robust security policy and a fault-tolerant mechanism, making it suitable for critical and high-confidence applications [3, 4]. In CMAF, any system is divided into three categories: real agent system (ReAS), replica agent system (RAS), and proxy agent system (PAS). ReAS acts a master system which can create a new agent and invoke any agent in the network for any task and also send an agent to other systems in the network. RAS is used for fault tolerance; it acts as ReAS in case current ReAS fails due to any reason. All other systems are considered as PAS.

Fig. 1 Multi-agent framework

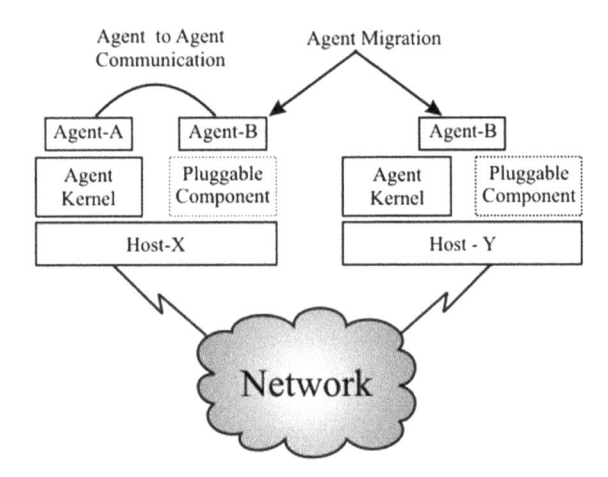

2.2 SCADA Testbed Architecture

SCADA testbed is modeled in defense-in-depth architecture which is a concept of layering multiple security solutions, so that if one is bypassed, another will provide the defense. Defense-in-depth is realized through the use of multi-agent-based implementation. Defense-in-depth architecture strategy includes the use of firewalls, digital simulator for simulating field devices, RTUs, detection and analysis of threats and vulnerabilities, and incident response mechanisms. In addition, an effective defense-in-depth strategy requires a thorough understanding of possible attack vectors [5, 6] which include backdoors and holes in network perimeter, vulnerabilities in protocols, attacks on field devices, database attacks, communications hijacking, and man-in-the-middle attacks.

The experimental testbed is shown in Fig. 2. This environment is used to reproduce the dynamics of SCADA system along with additional facilities and mechanisms for simulation and impact analysis in a safe, systematic, and rigorous way. CMAF is used in the testbed for simulating attack scenarios, impact analysis of attacks.

SCADA testbed is composed of field network, control network, process control network, demilitarized zone, corporate network, external network. Testbed uses Central Power Research Institute's real-time digital simulator for simulation of field devices. Control network acquires the data from field devices using remote terminal unit (RTU) as well as RTU simulators and sends to process control network. Process control network is comprised of master terminal unit (MTU), human–machine interface (HMI), data historian, etc. The link between MTU and RTUs may be point to point or point to multi-point. IEC 60870-5 and ICCP protocols are used as communication protocols [7]. HMI gets the data from the MTU and displays the data using mimics. Database server acquires the data from master station and stores for analysis. Demilitarized zone is used to provide an extra layer of

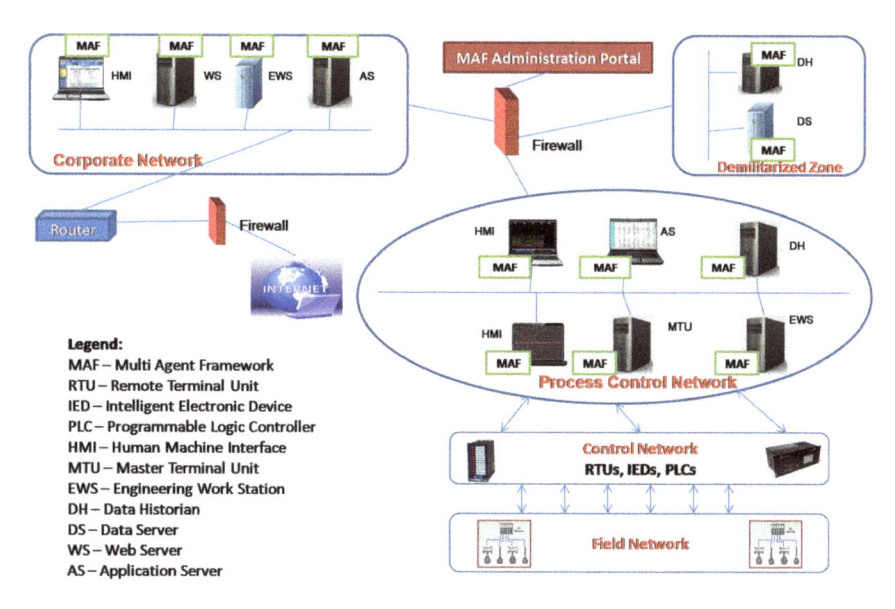

Fig. 2 Proposed SCADA testbed architecture

security by separating the corporate network and the SCADA network and also introduces additional firewall. Corporate network is having various servers such as Web servers, application servers, and human–machine interface. These servers acquire the data from the data server and use for different purposes. External network is any kind of public network, e.g., Internet. At the moment, it is quite common to use Internet for connecting different industrial installations belonging to the same company. These connections are the principal source of attacks against SCADA. For this reason, we simulated this testbed in our isolated environment.

Deployment of Multi-Agent Framework: Multi-agent framework is installed in every system in process control network, demilitarized zone, and corporate network. Among all systems, one system acts as a ReAS, one system acts as RAS, and other systems act as PAS. ReAS will be coordinating with all agent systems for info collection and installation. RAS will be used for fault tolerance, where this system acts as ReAS in case of current ReAS fails to work. All other systems will be treated as PAS, where agent framework will be running. This framework is provided with administration portal for controlling the testbed as well as for simulating various attacks.

Tools for testbed: SCADA threat analyzer (STA) and security information and event management (SIEM) tools are developed for simulation and impact analysis of attacks on testbed. STA tool can be used for simulation of attacks and analysis. STA provides a dashboard for system-wide threat analysis. SIEM tool provides security information about every system in testbed over a dashboard. Both tools use CMAF as a base framework.

3 Simulation and Impact Analysis of Attacks Using Testbed

For proposing countermeasures for various attacks, it is required to simulate the attacks to understand the methodology of attack. Using the testbed, we can implement specific experimental attack scenarios that compromise the integrity, confidentiality, and availability of the entire system. Analyzing each type of attack regarding these three characteristics will make it easier to identify the consequences of each attack [8]. Here we list some of the attacks simulated on our testbed.

The specific experiment scenarios that can be analyzed are internal attacks and external attacks. Any attack can be simulated by configuring CMAF deployed on the testbed. Denial of service, malware and the man-in-the-middle attacks are the major attacks [9, 10] in the SCADA system.

Denial-of-Service Attacks: The objective of DoS attack is to consume all network bandwidth and resources of the victim's machine. This attack can be carried out by targeting three different layers, viz. network layer, transport layer, and application layer. The network bandwidth of the victim's machine can be consumed by flooding random IP packets at network layer, and the resources can be consumed by either flooding SYN packets at transport layer or continuous user data packets at application layer. The experimental results of the three different DoS attacks are shown in Figs. 3, 4 and 5 [11].

Man-In-The-Middle (MiTM) Attack: The usage of open standard protocols for communication among SCADA components that are not designed with security in mind is one of the vulnerable areas for any attacker. The SCADA components in the control region, viz. the MTU and the RTU, communicate via these insecurely designed open protocols without any authentication. An attacker can target this communication and can launch an attack like the man-in-the-middle attack resulting in a disastrous situation. An experiment was conducted by exploiting the

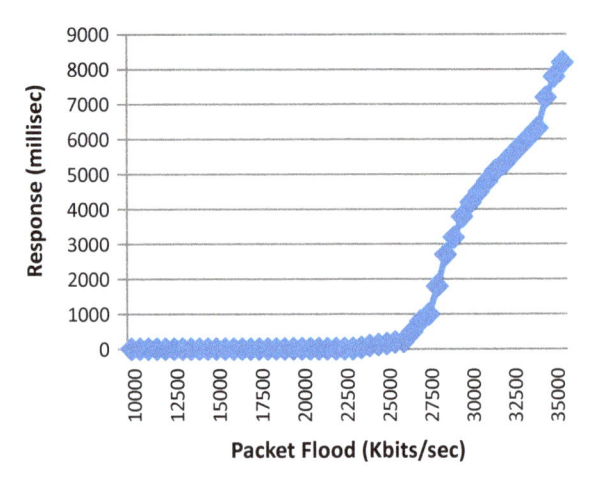

Fig. 3 IP packet flood versus response time

Fig. 4 TCP SYN flood versus response time

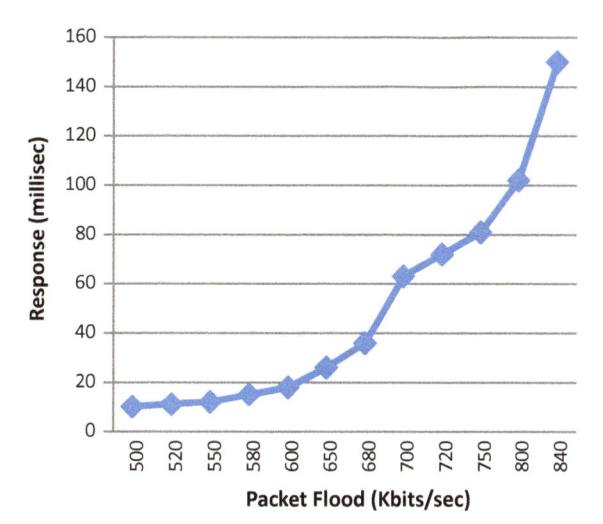

Fig. 5 IEC 60870-5-104 application protocol packet flood versus response time

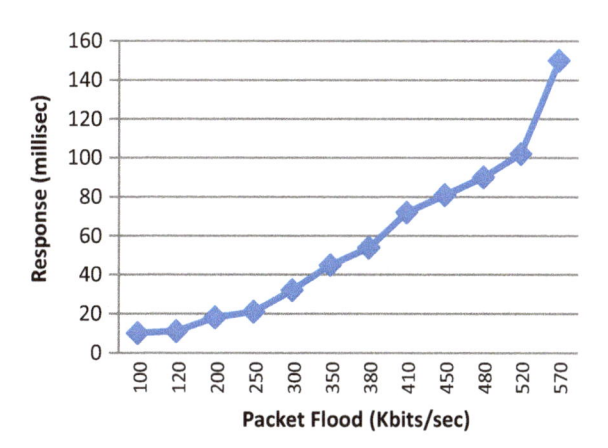

vulnerabilities of IEC 60870-5-104 at Central Power Research Institute, Bengaluru, using real-time digital simulator (RTDS) by simulating Salem substation to study the impact of an attack as shown in Fig. 6, and it was observed that the substation was destabilized [12].

Malware Attack: The attack of malicious software was conducted through mobile agents over SCADA testbed. Mobile agent (CMAF) was chosen for the attack environment because it enabled easier communication over SCADA network through code migration and node-to-node communication which was essential for analyzing malware characteristics and behavior. Also the worms could also be analyzed using mobile agent in easier manner as agent can propagate over the SCADA network in autonomic way, i.e., self-governing or taking decisions on its own. The malware can be injected into SCADA network using the SCADA threat analyzer. The next section discusses the malware attacks and analysis in detail.

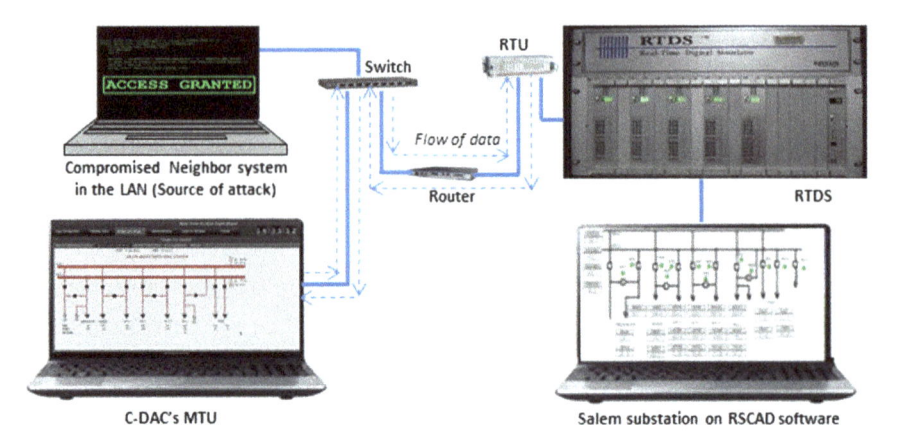

Fig. 6 Experimental setup at CPRI

4 Case Study: Impact Analysis of Malwares Using STA and SIEM Tools

Various methods are devised to detect, predict, and quantify the impact of security threat attacks on the SCADA system. An exhaustive analysis of all possible attacks is not feasible, but attack trees are generally used in the literature to categorize different types of attacks [13].

Various attack vectors such as network communication, protocols, system usage need to be analyzed to detect the attack on SCADA systems [14]. STA and SIEM tools are used over a SCADA testbed for analysis of any malware through characteristics and security information extraction.

4.1 Analysis using a SCADA Threat Analyzer (STA)

STA will provide the facility to perform experiments over the SCADA Testbed. A single experiment encompasses of single malware injection and analysis through characteristics extraction and presentation over a Web-based interface. The analysis can be performed in two different modes. First, malware with known characteristics can be injected into the SCADA testbed network using the threat analyzer tool and analyzed by extracting the characteristics and behavior of the malware. Second, a new malware can be injected with unknown characteristics and then analyzed for the characteristics.

STA is capable of 'worm tracing' also, injecting on one system and tracing on all the systems in the network. For every experiment, log files are generated having characteristics of malware. The process of analyzing malware characteristics is done in cycles with fixed time interval (sleep time) between two cycles of analysis.

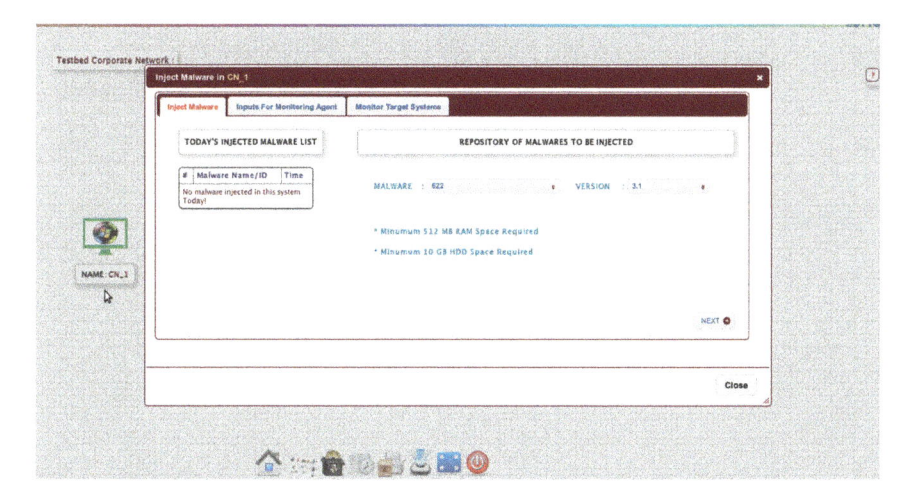

Impact Analysis of Attacks Using Agent-Based SCADA Testbed 49

Fig. 7 Injecting the malware over STA

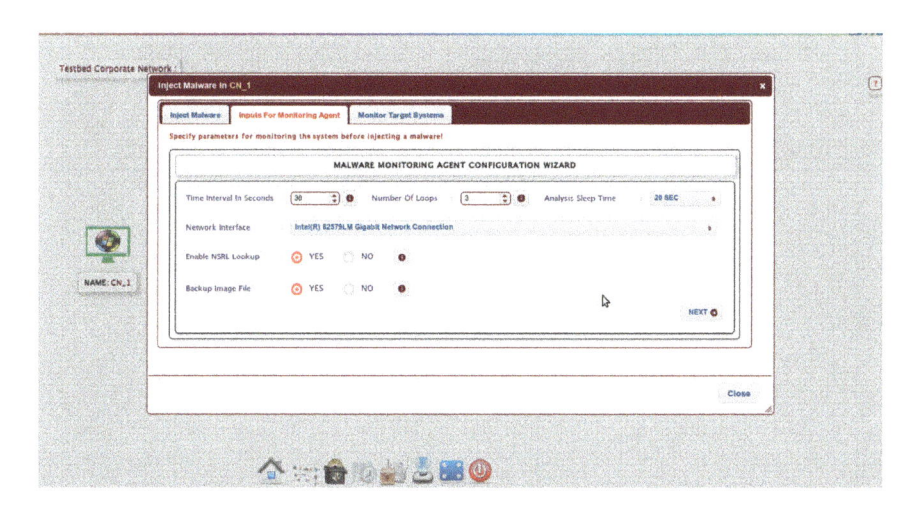

Fig. 8 Input parameters for injecting the malware over STA

This is done as some malwares may be active for some time and dormant for some time. This cyclic analysis can be configured using STA for any single experiment. The following screenshot in Figs. 7 and 8 shows the malware injection details.

The analysis under SCADA threat analyzer can be performed in two ways, namely baseline analysis and detailed analysis. Baseline analysis analyzes various characteristics through signature scan, process, file, services and registry changes, nsrl lookup for the complete system. Figure 9 shows the snapshot of baseline analysis.

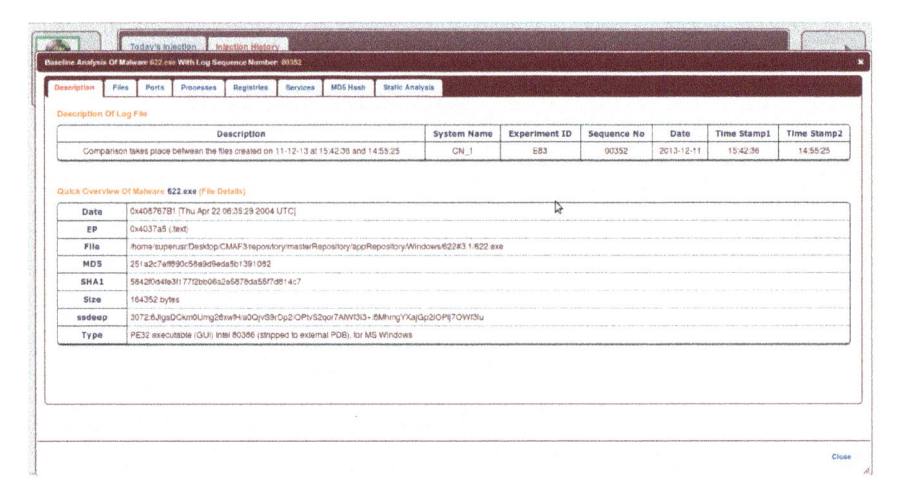

Fig. 9 Baseline analysis over STA

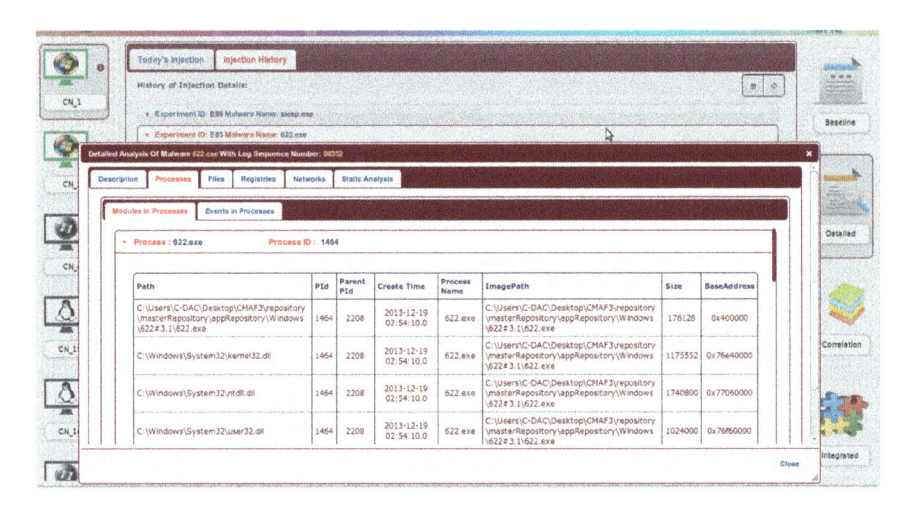

Fig. 10 Detailed analysis over STA

Detailed analysis extracts all the characteristics for the process or processes initiated by the malware sample by identifying the malware sample through its MD5 hash. Figure 10 shows the snapshot of detailed analysis.

The various features provided in STA include base-lining systems, complete system monitoring, memory analysis, packet capture, nsrl lookup, and dashboard-based forensics reporting/analysis.

Base-Lining Systems: Before injecting or simulation of attack over the SCADA systems, snapshot of the state of the systems is taken, which considers what all processes are running, what all system files are available, which ports are open, which services are running, and registries information, etc. During the monitoring of characteristics, every state is compared with the baseline state during the first cycle of active malware.

Complete System Monitoring: File, process, service and registry being created, destroyed or modified during the active cycle of the injected malware as compared with the previous cycle or with the baseline for first cycle are monitored. Further, signature or MD5 hashes of the files are scanned for modification during a cycle. Also the network statistics like which ports are closed or opened during the active malware cycle is monitored.

Memory Analysis: The volatility framework is used for memory analysis which is a completely open collection of tools, for pulling out of digital artifacts from volatile memory (RAM) samples [15]. The memory analysis is performed completely independent of the system being investigated but presents an insightful user interface for analyzing the various sections of the RAM activities showing runtime state of the system.

Packet Capture: Network activity is volatile and dynamic. Network forensics therefore requires a proactive approach to capturing network data. Here forensic network data is captured for every active malware cycle and presented in understandable format over a Web-based interface which can be further interpreted for tracing data leakage or detecting advanced persistent threats.

NSRL Lookup: The National Software Reference Library (NSRL) is designed to collect software from various sources and incorporate file profiles computed from this software into a Reference Data Set (RDS) of information. The RDS is a collection of digital signatures of known, traceable software applications [16]. The threat analyzer developed by C-DAC uses the existing tools under the NSRLquery project. Two nsrlsvr are run simultaneously for lookup, one is the default nsrlsvr instance consisting of NSRL RDS, and another instance is customized to contain the digital signatures of the software's run by the threat analyzer and is known to be good.

nsrllookup is run on the proxy system or vulnerable system to lookup for any files created which are not falling within the known files category specified in either of the nsrlsvr RDS. The nsrllookup uses the MD5 hash for matching with the existing known hashes and gives the unknown files as the output which may be considered malicious [17, 18].

Dashboard: STA is provided for showing the features mentioned and comparison results in easy-to-analyze and organized way.

4.2 Analysis using a Security Information and Event Management (SIEM) Dashboard

SIEM dashboard provides the insight into the activity in each and every system of the SCADA testbed over a single window. This would help the operator to analyze

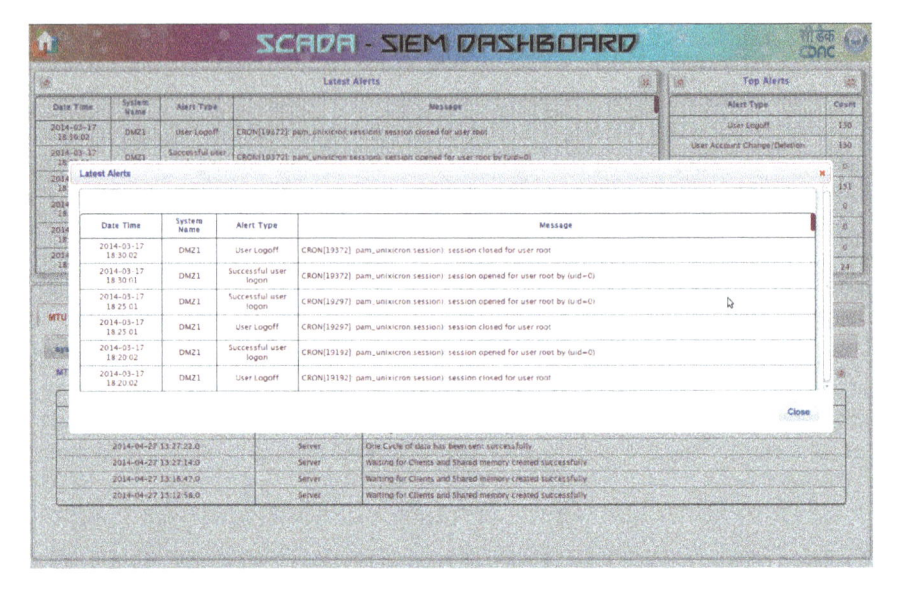

Fig. 11 SIEM dashboard

any misbehavior of any system in faster and easier way through appropriate alerts messages [19, 20].

SIEM tool is a Web-based interface for viewing the different application and system logs generated by any system in the SCADA network over a single dashboard. The dashboard provides system-wise real-time alerts, latest alerts, and top alerts for monitoring the system or application logs of every system in the SCADA network. The dashboard view is as shown in Fig. 11.

System-Wise Real-Time System/Application logs: This section displays the instantaneous details of system or application logs generated for a particular system in the SCADA network like an MTU is shown in the Fig. 11. The 'message level' column displays as 'information' or a 'warning' which is used as alert message to take appropriate action for the operator.

Latest Alerts: This section displays the details of the any system in the SCADA network in a scrolling fashion and sorted based on time of event. The 'Alert type' column displays 'user logoff,' 'successful user logon,' etc., which can be used as indicator for any timely action by operator.

Top Alerts: This section displays the no of occurrences of a particular event like 'Password Changes,' 'Service started,' 'user logoff,' or 'successful user logon' indicated through column 'Alert type' which can be used for any timely action by operator if the no of suppose 'successful user logon' is increasing abruptly. The possible reason for any increase in a particular alert message like 'successful user logon' may be found to check the exhaustion of resources and making systems unavailable.

5 Conclusion

The paper presents an experimental testbed for any SCADA system to be analyzed for various attack scenarios. Some of the major attacks over SCADA system like DoS, MiTM, and malware attacks were simulated over the testbed, and the impact has been analyzed. The SCADA testbed is equipped with a multi-agent framework for any communication across the systems in the network. STA and SIEM tools are used for simulating attacks over testbed and analyzing the impact through characteristics and security information extraction. Malware attack over SCADA testbed is used as the case study for impact analysis of the attack over SCADA System. The analysis results derived are found to be matching with the results of any other malware analysis tool. The malware attack analysis mentioned in the paper provides additional features such as extensive memory analysis, nsrl lookup which may not be available in any other malware analysis tool which use a sandboxed environment for analysis and can be evaded by malwares [21].

References

1. Amanullah MTO, Kalam A, Zayegh A (2005) Network security vulnerabilities in SCADA and EMS. In: Transmission and distribution conference and exhibition: Asia and Pacific, 2005 IEEE/PES. IEEE
2. Queiroz C et al (2009) Building a SCADA security testbed. In: Third international conference on IEEE network and system security, 2009 NSS'09. IEEE
3. Venkatesh S, Bindhumadhava B, Bhandari A (2006) Implementation of automated grid software management tool: a mobile agent based approach. IKE
4. Raghu HV, Saurav SK, Bapu BS (2013) PAAS: Power aware algorithm for scheduling in high performance computing. In: Proceedings of the 2013 IEEE/ACM 6th international conference on utility and cloud computing, IEEE Computer Society
5. Top 10 Vulnerabilities of Control Systems and Their Associated Mitigations (2007) North American electric reliability council control systems security working group
6. Weiss J (2008) Key issues for implementing a prudent control system cyber security program, Electric Energy T&D Magazine (March–April 2008)
7. Mohagheghi S, Stoupis J, Wang Z (2009) Communication protocols and networks for power systems-current status and future trends. In: Power systems conference and exposition, 2009. PSCE'09. IEEE/PES. IEEE
8. Giani A et al (2008) A testbed for secure and robust SCADA systems. ACM SIGBED Rev 5(2):4
9. Long M, Wu C-H, Hung JY (2005) Denial of service attacks on network-based control system: impact and mitigation. IEEE Trans Ind Inf 1(2):85–96
10. Davis CM, Tate JE, Okhravl H, Grier C, Overbye TJ, Nicol D (2006) SCADA cybersecurity test bed development. In: Power symposium, NAPS 2006, pp 483–488
11. Kalluri R, Mahendra L, Senthil kumar RK, Ganga Prasad GL (2016) Simulation and impact analysis of DoS attacks on power SCADA at NPSC 2016, IIT Bhubaneswar
12. Abhiram A, Mahendra L, Kalluri R, Senthil kumar RK, Ganga Prasad GL (2015) Transient analysis of cyber-attacks on power SCADA using RTDS. J CPRI 11(1):77–80
13. Byres EJ, Franz M, Miller D (2004) The use of attack trees in assessing vulnerabilities in SCADA systems. In: Proceedings of the international infrastructure survivability workshop

14. Zhu B, Joseph A, Sastry S (2011) A taxonomy of cyber attacks on SCADA systems. In: Internet of things (iThings/CPSCom), 2011 international conference on and 4th international conference on cyber, physical and social computing. IEEE
15. The Volatility Foundation—Open Source Memory Forensics. http://www.volatilityfoundation.org/
16. Rowe NC (2012) Testing the national software reference library. Dig Investig 9:S131–S138
17. National Software Reference Library (NSRL) NSRLquery project. http://www.nsrl.nist.gov/
18. Spreitzenbarth M, Uhrmann J (2015) Mastering python forensics. Packt Publishing Ltd
19. Coppolino L et al (2011) Integration of a system for critical infrastructure protection with the OSSIM SIEM platform: a dam case study. In: International conference on computer safety, reliability, and security. Springer, Berlin
20. Coppolino L et al (2012) Enhancing SIEM technology to protect critical infrastructures. In: International workshop on critical information infrastructures security. Springer, Berlin
21. Lindorfer M, Kolbitsch C, Milani Comparetti P (2011) Detecting environment-sensitive malware. In: International workshop on recent advances in intrusion detection. Springer, Berlin

Power Exchange and Its Significance to Enhance the Deployment of Smart Microgrids in India and Key Barriers in Its Adoption

Furkan Ahmad, Mohammad Saad Alam, Reena Suri,
Akhilesh Awasthy and M. Shahidehpour

Abstract Evolution of power sector in the last decade was concomitantly sustained by formation of institutions that enhance efficiency through markets by bilateral trading which soon after 2008 through trading on power exchanges (PXs). The Electricity Act (EA) 2003 unbolted the power sector by laying down provisions for promoting competition in the power market. This paper presents a comprehensive framework of PXs-based electricity trading market (comprises of DAM, TAM, RECs, and ESCs), PXs significance in order to enhance the deployment of renewable energy sources (RESs)-based smart microgrids in Indian scenario, and key barriers in larger scale adoption. Further, to validate the significance of PXs-based market, a case study of solar PV-based smart microgrid has been executed, which reduces 40.79% net payable amount as compared to DISCOM, and 32.05% through bilateral per year.

Keywords Smart microgrid · Indian power market · DISCOM
Bilateral trading · Power exchange · Day-ahead market

F. Ahmad (✉)
Department of Electrical Engineering, Aligarh Muslim University,
Aligarh 202002, India
e-mail: furkanahmad@zhcet.ac.in

M. S. Alam
Center of Advanced Research in Electrified Transportation (CARET),
Aligarh Muslim University, Aligarh 202002, UP, India
e-mail: saad.alam@zhcet.ac.in

R. Suri
India Smart Grid Forum, New Delhi, India
e-mail: reena.suri@indiasmartgrid.org

A. Awasthy
Indian Energy Exchange Limited, New Delhi, India
e-mail: akhilesh.awasthy@iexindia.com

M. Shahidehpour
Electrical and Computer Engineering Department,
Illinois Institute of Technology, Chicago, USA
e-mail: ms@iit.edu

© Springer Nature Singapore Pte Ltd. 2018
R. K. Pillai et al. (eds.), *ISGW 2017: Compendium of Technical Papers*, Lecture Notes in Electrical Engineering 487, https://doi.org/10.1007/978-981-10-8249-8_5

1 Introduction

Electrical energy and its convenient accessibility are a bit up a benchmark of advancement and a key essential for all-circular prosperity. No core financial pursuit can be sustained without adequate and consistent source of energy. In India, the total RES-based power generation installed capacity is 42,849.38 MW (grid connected) and 1329.76 MW (off grid); i.e., 13% of total installed capacity has been reported by the Ministry of New and Renewable Energy (MNRE) till June 30, 2016. However, 70% of total installed capacity comes from fossil fuel-based generating station (61% coal and 9% natural gas), which rises environmental issues [1]. In concern with increasing demand and GHG emission, renewable energy recourses (RESs)-based microgrid deployment is gaining massive attention due to various benefits associated such as uncontaminated energy, zero carbon emission, free from dependency on foreign oil which also enhance economic infrastructure. As being smart in nature microgrids are capable to operate in both modes, i.e., grid connected (grid tied/grid interactive) as well as in islanding mode [2–4].

The deployment of microgrids on small scale enhances the single entity system such as a home, school, or particular load near to the distributor generators (DGs). In order to meet the increasing demand of electrical power and overcome/control greenhouse gas emission (GHG), the production of uncontaminated green energy must be introduced from each corner of the country, but due to dependency on metrological condition it is not feasible; in such state of affairs, the RES-based power production at large scale and its transmission through the power market is only viable solution [5].

Indian power market has been highly complex entity due to separate electricity regulation commissions (ERC) of each state, and most of them follow the vertical integrated market. After the implementation of Electricity Act 2003, open access market has been introduced to end up the monopolies and enhance the liquidity in electricity power market. The Indian power market is divided into three type of contracts: long term that follow up by power purchase agreement (PPA), medium-term contracts, and short-term contracts. After inception of power exchanges (PXs) market in 2008, the efficiency and competition in power market improved and volatility decreased [6–10].

Rest of the paper is organized as follows: Sect. 2 deals with Indian power market, power trading, i.e., bilateral/open access. Section 3 deals with details of PXs-based power market which comprises of day-ahead market (DAM), term-ahead market (TAM), renewable energy certificates (RESs), and energy savings certificates (ESCs). Section 4 deals with the renewable energy status and key barrier in their adoption. Section 5 presents a case study to validate the significance of PXs to enhance the deployment of RESs-based smart microgrid in India.

2 Power Trading Framework in India

2.1 Power Trading

Power trade fundamentally implies a trade of power to an individual buyer with an option of price conciliation and power quantity. In India, power trading is sprouting with limited volumes of exchange. The consumer's demands of electrical energy are principally functioned by their corresponding government utilities like State Electricity Boards (SEBs) or their beneficiary entities, power departments, private licenses, etc., creating the monopoly with a single supplier. Similar scenario is in the generation of electricity as Central Generating Stations (CGSs), Independent Power Producers (IPPs), and SEBs have all their capacities tied up with each SEBs having assigned share in central projects. In other words, the power suppliers have a minute option about whom to trade the power and the purchasers have no option between power providers [11].

The power rate has chiefly been fixed/controlled by the Central/State Governments or by the Regulatory Commissions at the Centre/States wherever they are already functional. Power generation/transmission is highly capital intensive, and the fixed charge component makes up a major part of the tariff. India has a largely agricultural economy, heavily populated seasonal, and per hour power utilization of the day with variants in peak hours and off-peak hours. Moreover, the geographical diversity of India has different climatic regions and different types of loads [6–10].

The exchange of electrical power from surplus State Utilities for deficit ones, through marginal investment, is eliminating grid constraints and also helps in conceding or dropping the investment for additional generation capacity. More preciously, the scheduled exchange of power will increase and unscheduled exchange will reduce bringing in grid discipline, a familiar problem. Power trading has been categorized in various forms such as bilateral, OTC, and power exchange-based open access power market [6–10].

2.2 Open Access Power Market

Alternately, open access is defined as enabling of nondiscriminatory trade system of electric power between two entities utilizing the state/central existing system (i.e., transmission lines) of an in-between third party, and not blocking it on unreasoning states. Interstate open access is such kind of power market where the seller and buyer parties belong to the different state, but the bid area can be same, while vice versa for intrastate open access. Detail of power trading structure for are shown in Table 1 [12, 13].

Table 1 Detail of trading duration

Type of market	Interstate	Intrastate
Short term	Intraday to 1 month	Designed by respective SERC
Medium term	3 month to 3 year	
Long term	12 year to 25 year	

2.3 Transactions in Open Access

There are various obligations to be remunerated by open access trades and customers to distribution licensee, transmission licensees, and other associated entities, other than the power procured cost paid for the generator or supplying entity. These obligations also comprise of connectivity charges, PoC charges, transmission charges and losses, wheeling charges and losses, cross subsidy surcharge, SLDC charges, and RLDC charges [12, 13].

3 Trading on Power Exchanges (PXs)

The efficiency and liquidity of the PXs have upgraded since their initiation in 2008. This is revealed by the reduction in a number of price peaks, volatility, and reflection of market information in prices. The size of PXs-based markets has grown up to approximately 3% of the total electricity generated at the moment. With further tightening of the frequency band and an expected zero tolerance toward frequency deviation, the sizes are expected to move from the real time to DAM or intraday electricity markets on the PXs. The cumulative percentage of electricity traded through the UI and PXs today stands at approximately 6%. Indian PXs comprises of four kinds of trading: Day-Ahead Market (DAM), Term-Ahead Market (TAM), Renewable Energy Certificates (RECs), and Energy Saving Certificates (ESCs). The total traded electricity in billion unit (BU) at PXs from last 2008 to 2016 has been given in Fig. 1, which shows the continuous increment of power trading at PXs. The yearly average per unit discovered price is shown in Fig. 2, which indicate that due to rising of competitivity and liquidity the per unit price is continually falling. In the starting year of PXs during 2008–09, 7.49 INR/kWh was recorded, while in 2015–16 it was only 2.72 INR/kWh [11, 14, 15].

3.1 Day-Ahead Market (DAM)

An electricity open access market where trading takes place for the next day based on double side closed bidding system. The delivery of electricity starts from

Fig. 1 Total transactions at PXs from FY 2008–09 to 2015–16

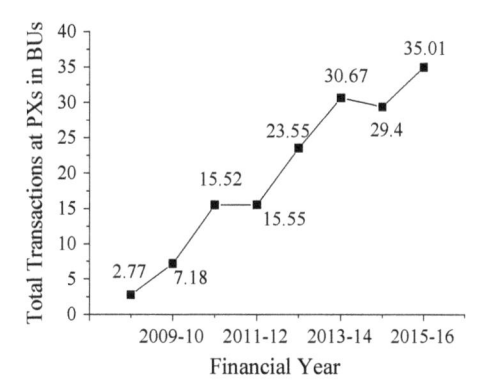

Fig. 2 Yearly avg. discovered price at PXs (INR/kWh)

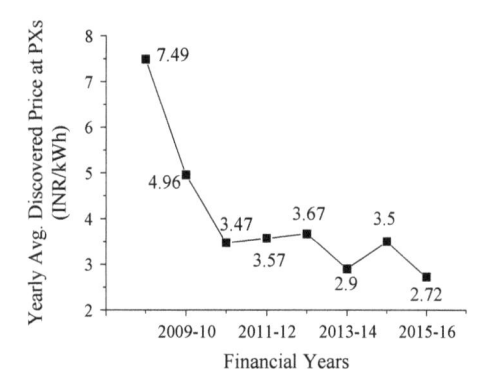

midnight on the bidding day till next 24 h. In India, DAM has started since September 2008. CERC open access regulations have been followed and scheduling is governed by Power Grid Corporation of India Limited (PGCIL) [11].

The DAM encompasses of various steps such as bidding, matching, transmission corridor and funds availability, results, confirmation and scheduling.

For DAM, bidding starts from 10:00 to 12:00 h. The seller and buyer can bid for electricity in terms of 15-min blocks, minimum one block to maximum 96 blocks. The amount of power for which a client/member may bid is 0.1–50 MW; maximum possible bidding price will be not more than INR 20/kWh. Further bidding strategy is divided into three categories.

Single bids system, where seller or buyer can bid for a single 15-min block or a set of blocks or random blocks but maximum 96 blocks with the different/same price. All the quantum of bidding under the market clearing price (MCP) will be accepted; i.e., partial execution is possible; in block bids system, a seller/buyer can bid for a single 15-min block or a series of 15-min blocks. Entire bidding quantum of the "seller/buyer" will accept or reject depending on the average of entire quantum for which a buyer bid is equal or greater than the MCP, and in mother–child bid system, the acceptance of bidding of one or set of one quantum depends

Table 2 Timeline for DAM [16]

Process	Schedule
Initial margin check	09:30 AM
Bidding session • Double side closed bidding (single or block), each block 15 min • Member/client may edit, modify/delete buy and sell bids	10:00 AM to 12:00 AM
Calculation of MCP < MCV and provisional obligations of the members/NOC from the states Communication with NDLS for unconstrained solution Confirmation from the bank regarding the funds availability in member/client accounts and	12:00 AM to 01:00 PM
NDLS checks the congestion and deals with the situation	01:00 PM to 02:00 PM
Buyers pay to PXs (pay-in)	By 02:30 PM
Calculation of area clearing prices (ACP) based on transmission available and communicate NLDC for scheduling	By 03:00 PM
NLDC confirms scheduling. PXs send a detailed schedule to SLDCs	By 05:30 PM
RLDCs/SLDCs incorporate collective transactions in the daily schedule	By 06:00 PM
PXs settled payments to the seller (playouts)	Next day by 02:00 PM

upon the acceptance on previous quantum. The complete scheduling/timeline for DAM is given in Table 2.

After the bidding session for DAM, price calculation algorithm will determine the MCP and MCV from the intersection point of demand (buy bid) and supply (sell bid) curve. The intersection point of both curves represents the price at which the total sell quantum and buy quantum are equal. For better understanding, an example is shown in Table 3 and Fig. 3, where entities 1, 2, and 3 participated in DAM and each entity bided for different quantum at different price. At price INR 2.5, the sell and buy both quanta are equal. So this price will be MCP and respective volume will be MCV [16].

3.2 Term-Ahead Market (TAM)

A kind of open access power market where member/client may sell/buy electricity on the term basis for a maximum duration of 11 days from the same day. In the Indian power market, "TAM" was incorporated in September 2009 under CERC Open Access Regulations 2008.

TAM comprises of four contracts: Intraday, Day-Ahead Contingency (DAC), Daily and Weekly Contracts. Both weekly and daily further have four contracts: Round the Clock (RTC for 24 h), Day (for 11 h, 7 AM to 6 PM), Night (for 8 h, 11 PM to & AM), and Peak (for 5 h, from 6 PM to 11 PM). The timeline is shown in Table 4 [17].

Table 3 Entity 1, 2, and 3 participated in DAM

Price tick (Rs.)	Entity 1 (MW)	Entity 2 (MW)	Entity 3 (MW)	Total buy quantum (MW)	Total sell quantum (MW)	Net transaction (MW)
1.0	−10	40	40	80	−10	70
1.2	−20	50	70	120	−20	100
1.5	40	60	−20	100	−20	80
2.1	44	60	−20	104	−20	84
2.3	30	50	−30	80	−30	50
2.5	**20**	**40**	**−60**	**60**	**−60**	**00**
2.9	20	45	−60	55	−60	05
3.4	00	42	−80	42	−80	−38
4.7	00	41	−85	41	−85	−44
5.2	00	50	−105	50	−105	−55
5.9	00	20	−110	20	−110	−90
6.3	00	20	−120	20	−120	−100
7.6	00	20	−100	20	−100	−84
8.0	00	20	−100	20	−100	−80

Fig. 3 Intersection of supply–demand curve to determine MCP and MCV

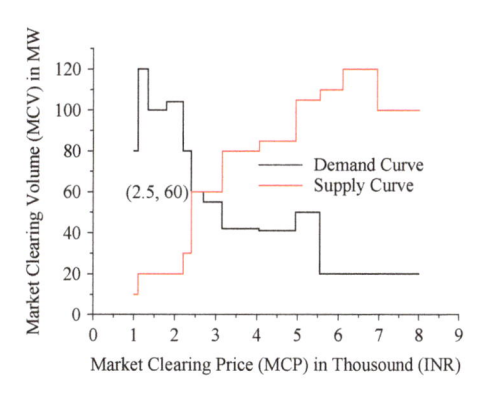

3.3 Renewable Energy Certificates (RECs)

The CERC introduced REC mechanism to abridge the procurement of renewable energy by the state as well as private utilities and obligated entities, including the states which are not well endowed with RESs. REC framework aims to create a national level market for renewable generators to recover their cost. One REC signifies 1 MWh of energy generated from RESs. Under the RECs mechanism, a producer can generate electrical energy through the RESs in any part of the nation. For the electricity part, the producer obtains the cost equivalent to that from any

Table 4 Time for TAM [17]

Features	Intraday	DAC	Daily	Weekly
Duration	Same day, hourly, for 10 h	Hourly for next day	All or hourly in a single day	Monday to Sunday
Trading day	Every day			Wednesday and Thursday
Delivery period	02:00 PM to 12:00 PM	RTC for next day	4th day to next 7 days	Max 11 days
Trading time	10:00 AM to 05:00 PM	03:00 PM to 05:00 PM	12:00 AM to 03:00 PM	12:00 AM to 04:00 PM
Bid matching	Continuous trading			Double side open auction system
Delivery point	Regional periphery			
Pay-in	Delivery day −1			
Payout	Delivery day +1			
Transaction fee	2 paisa/kWh, fixed by CERC			

Table 5 Time line for RECs [19]

Trading process	Schedule
Trading day	Last Wednesday of each month
Market type	Closed double-sided auction
Trading schedule	01:00–03:00 PM
By 15:30 PM	Validation of REC by central agency
By 04:00 PM	Confirmation from central agency
By 04:30 PM	PXs finalize trade
By 05:00 PM	Buyers and sellers informed to the central agency
By 06:00 h	Invoice raised

conventional (traditional) source while the environment attribute is sold through the PXs at the market-determined price. The obligated entity from any part of the nation can procure these RECs to meet its RPO compliance [18].

One REC indicates 1 MW contribution in RESs generation with the validity of 1095 days (three years). The RECs has been divided into solar (INR3500/REC) and nonsolar (INR 1500/REC), but the forbearance price for both types of REC is INR 5800 and INR 3300. In RECs trading, banking and borrowing, both are not allowed. The timeline for RECs trading at PXs is given in Table 5 [19].

3.4 Energy Savings Certificates (ESCs)

To enhance the energy saving, Perform, Achieve and Trade (PAT) mechanism was launched by Bureau of Energy Efficiency (BEE) on July 2012 under the effect of the ministry of power. ESCs are referred as a white certificate. A total of 478 designated consumers (DCs) have been covered from eight power oriented sectors—thermal power, aluminum industry, iron and steel, cement, pulp and paper, fertilizer, chloralkali, and textiles. All these sectors contribute 25% of GDP and also accountable for approximate 45% of India's primary energy consumption [20].

4 Renewable Energy and Key Barriers in Its Adoption

The Indian government has targeted 175 GW installed capacity based on RES (100 GW solar-based and 60 based on wind energy and 15 rest of RES) by 2022 [21]. Numerous roadblocks causing slower growth rate of RESs technology adoption in India depend upon the several factors; some of them are identified here such as high capital cost, lack of financing mechanism and governmental subsidies, transmission and distribution losses and grid complexity, lack of public interest: consumer awareness and insufficient market base, lack of paying capacity, undeveloped backup or storage device, unavailability of solar radiation data, lack of weather, and load forecasting techniques, lack of information/communication technology resources, less efficient technology, lack of experienced expertise and R&D technical institutes, lack of local infrastructure as well as national infrastructure, scarcity of natural and new and renewable resources, geographic conditions and ecological issues, rehabilitation controversies, faith and beliefs toward RESs, lack of political commitment, and lack of adequate government policies toward electrical power generation from RESs [22].

5 Case Study

5.1 Load Profile Analysis

To determine the significance of PXs platform in order to enhance the deployment of smart microgrid in India, a variable load system was taken. The load system comprises of a community (homes, hospital, light street), a school, a hostel, and a small industry. The analysis of load was carried out on hour basis for financial year 2015–16. All the data was taken from the meter reading. Table 6 shows the peak power, average power, the average energy consumption, and load factor for each category of the load is given.

Table 6 Load description

Load	Average energy (kWh/day)	Average demand (kW)	Peak demand (kW)	Load factor
School load	343.2	14.3	77.8	0.18
Hostel load	1094.2	45.6	154.8	0.29
Community	15,984.9	666.0	1811.6	0.36
Industry	21,438.3	893.3	1532.1	0.58
Total load	38,860.74	1619.2	2288.6	0.53

5.2 Solar Radiation Data

The solar radiation data was taken from NASA meteorological center on an hourly basis for 365 days for the financial year 2015–16. Monthly average solar irradiance (5.67 kWh/m2/day) was recorded [23]. The monthly average daily solar radiation and corresponding clearing index are given in Fig. 4.

5.3 Power Trading Platform

RERC allowed open access trading in the generation as well as in consumer side. DISCOMs, bilateral, and PXs are the three trading platforms that are available to trade the power.

Figure 5 shows the trend in the DISCOM energy charges and feed-in tariff of Rajasthan state for last five financial years [24–26].

Figure 6 represents the trend in the IEX price rate at monthly average as well as selling and purchasing price after including all type of taxes, losses, and charges in

Fig. 4 Monthly average daily solar irradiance and clearing index

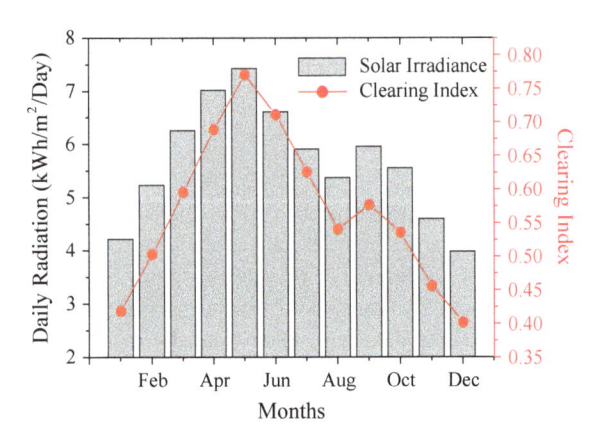

Fig. 5 Financial year-wise energy purchasing cost and solar feed-in tariff

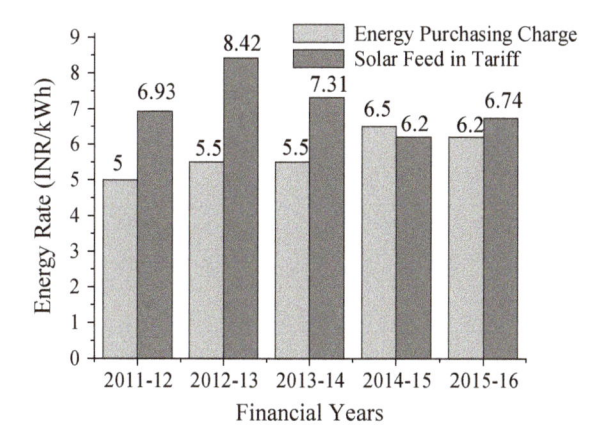

Fig. 6 Trend at IEX price rate for the financial year 2015–16

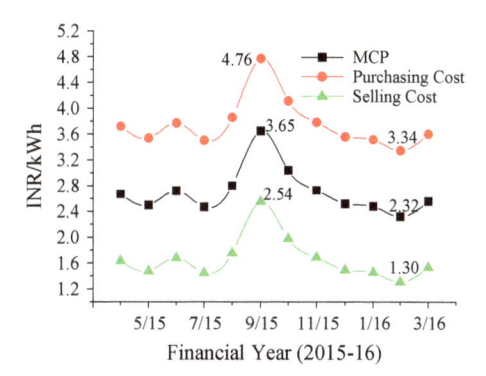

Fig. 7 Trend at bilateral price rate for the financial year 2015–16

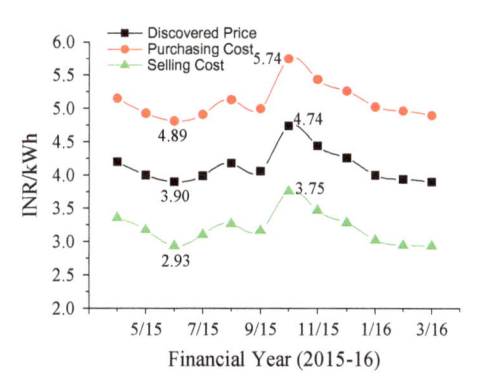

INR per kWh, where minimum 2.32 INR/kWh and maximum 3.6 INR/kWh were recorded as discovered price, while Fig. 7 shows INR per unit monthly average discovered price through bilateral trading for the financial year 2015–16. In bilateral market, minimum INR 3.9 and maximum 4.74 INR/kWh were recorded [10].

Fig. 8 Comparative study of power trading on various platforms

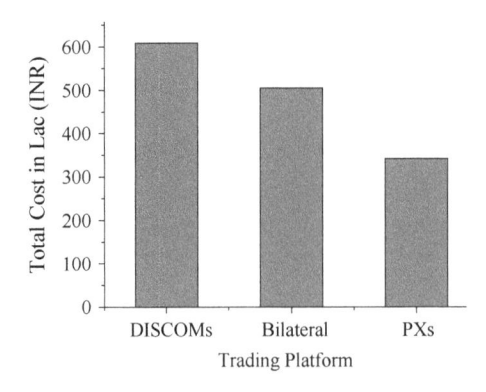

5.4 Results and Discussion

To determine the significance of PXs-based market, the surplus and deficit power on hourly quantum was traded through the various routes of the power market. On basis of power trading contracts, three cases are analyzed, which are as follows, power trading through DISCOM only, power trading through bilateral only, and power trading on IEXs (Under DAM). Power purchasing tariff and solar feed-in tariff cost at DISCOM was taken as INR 6.2 and 6.74 per kWh, and effective power sell and buy cost under bilateral and PXs has been discussed above.

After the analytical study of above three cases, it is observed that during the power shortages/deficit hour PXs is the most beneficial while for surplus hours DISCOM is most favorable. The prices of DISCOM, bilateral trade, and PXs are variable in nature. So, the efficient and economical operation of solar PV-based microgrid in power market depends on various factors such as variation in tariff, PXs price rate, and accurate solar and load forecasting for each hour.

Figure 8 shows the yearly net payable amount in lac; it reveals that the net payable minimum amount is 341.90 lac (INR) under PXs followed by bilateral trading contract 504.86 and 637.51 lac under DISCOM-based power market.

6 Conclusion

The mainstream of topical grid transformation efforts is directed toward the distribution systems to be able to meet new-fangled challenges in smart microgrid. It is one of the indispensable directions endorsed by policy architects for electricity delivery in various countries. The most significant contribution of this research deals with benefits of PXs-based open access market. In this paper, a comprehensive framework of power trading on PXs has been deliberated, taking into account numerous unique characteristics of this new marketplace. The details of various key barriers in adoption of RESs-based DGs are deliberated. To validate the significance of PXs-based market in order to enhance the deployment of RESs-based smart microgrid in India, a case study was executed. It reduces 40.79% net payable amount as compared to DISCOM, 32.05% trading through bilateral contracts per year. This research work establishes an economic rationale to the vision of the large-scale deployment of microgrids serving residential communities, industries, etc., in the near upcoming days and develops an ample understanding of the microgrid power market.

Acknowledgements This research work was supported by the Centre of Advanced Research in Electrified Transportation (CARET), Aligarh Muslim University, Aligarh, and Council of Scientific & Industrial Research (CSIR), New Delhi, India.

References

1. All India Installed Capacity in (MW) of Power Station as on 30-06-2016 (PDF). Central Electricity authority, GoI. Retrieved 11 July 2016
2. Parhizi S, Lotfi H, Khodaei A, Bahramirad S (2015) State of the Art in Research on Microgrids: A Review. IEEE Access 3:890–925
3. Liserre M, Sauter T, Hung J (2010) Future energy systems: integrating renewable energy sources into the smart power grid through industrial electronics. IEEE Ind Electron Mag 4(1):18–37
4. Spagnuolo G et al (2010) Renewable energy operation and conversion schemes: a summary of discussions during the seminar on renewable energy systems. IEEE Ind Electron Mag 4(1):38–51
5. Mohammadi M, Hosseinian SH, Gharehpetian GB (2012) Optimization of hybrid solar energy sources/wind turbine systems integrated to utility grids as microgrid (MG) under pool/bilateral/hybrid electricity market using PSO. Sol Energy 86(1):112–125
6. Karthikeyan SP, Raglend IJ, Kothari DP (2013) A review on market power in deregulated electricity market. Int J Electr Power Energy Syst 48:139–147
7. Bajpai P, Singh SN (2004) Electricity trading in competitive power market: an overview and key issues. In: International conference on power systems, ICPS 2004. Kathmandu, Nepal, pp 571–576
8. Shukla UK, Thampy A (2011) Analysis of competition and market power in the whole sale electricity market in India. Energy Policy 39:2699–2710
9. Joseph KL (2010) The politics of power: electricity reform in India. Energy Policy 38: 503–511
10. Indian Energy Exchange Annual Report 2015
11. Introduction to Indian Power Market: available at http://indianpowersector.com/home/about/overview/. Visited at 10 Sept 2016
12. http://indianpowersector.com/home/wp-content/uploads/2014/07/Report-OPEN-ACCESS-IN-INDIA.pdf
13. Power Trading/Open Access, Integrated Sustainable Energy Solutions. Available at http://powertrading.in/power-trading/
14. Trading market: available at http://indiapowertrading.info/
15. Report on Short-term Power Market in India: 2015–16. Economics Division Central Electricity Regulatory Commission. Available at http://www.cercind.gov.in/2016/MMC/AnnualReport15-16.pdf. Visited at 27 Aug 2016
16. Overview of Day Ahead Market. Available at http://www.iexindia.com/products.aspx?id=14&mid=2

17. Overview of term Ahead Market. Available at http://www.iexindia.com/Products.aspx?id=3&mid=2
18. Shrimali G, Tirumalachetty S (2013) Renewable energy certificate markets in India—a review. Renew Sustain Energy Rev 26:702–716
19. Renewable Energy Certificates. Available at http://www.iexindia.com/faqs.aspx?id=31&mid=1
20. Energy Saving Certificates. Available at http://www.iexindia.com/products.aspx?id=11&mid=1
21. Renewable energy's transformation of the Indian electricity landscape. Available at www.pwc.in/assets/pdfs/publications/2015/renewable-energys-transformation.pdf. Visited at 18 Sept 2016
22. Luthra S, Sanjay K, Dixit G, Abid H (2015) Barriers to renewable/sustainable energy technologies adoption: Indian perspective. Renew Sustain Energy Rev 41:762–776
23. Surface meteorology and solar energy: a renewable energy resource web site (release 6.0) sponsored by NASA's earth science enterprise program. Available at https://eosweb.larc.nasa.gov/sse/RETScreen/
24. 358-RERC Solar benchmark cost tariff. Available at http://www.ireeed.gov.in/policyfiles/358RERC%20Solar%20benchmark%20cost%20%20tarif%20FY14-15.pdf
25. Solar Tariff. Available at http://www.ireeed.gov.in/policyfiles/448CERC%20solar%20tariff.pdf
26. Rajasthan Electricity Regulatory Commission, Jaipur. Available at http://rerc.rajasthan.gov.in/TariffOrders/Order203.pdf

Low-Cost Spark-/Arc-Free Retrofit Smart Grid Switches Improve Distribution Quality and Reduce Distribution Losses Substantially

G. V. Sukumara and Vijay L. Sonavane

Abstract Smart power solution is the need of the hour for efficient power Distribution. Though abundant solar energy is being tapped, the absence of economical storage device has become a bottleneck to meet the peak power demand from 6 to 10 AM and 6 to 10 PM when solar energy cannot generate any power. In this situation, real-time control of power distribution has become necessary. Low-cost spark-/arc-free retrofit smart grid switches will be an efficient tool to implement this. These switches not only help for efficient grid management but also improve quality of power distribution and reduce distribution losses substantially. In urban and semi-urban power distribution due to continuous unbalance, the neutral current and losses are very high. Low-cost spark-/arc-free make before break transposing switches will reduce neutral current and the losses substantially. Many states have segregated the agricultural load. The probability of all the loads connected will be high when energized. The line gets loaded, and the voltage drops to as low as 300 V at the tail end. This not only increases losses but also results in burning of motors. Low-cost spark-/arc-free switches and time-based zone method of power distribution will not only reduce the distribution losses but also help to maintain voltage even at the tail end, reducing the probability of burning of motors due to low voltage.

Mrs. Vani Sukumar is the assignee of the process patent no 251143 dated 12/11/2004, and the patent for the product no 201641022530 with filing date of 30.06.2016 is pending.

G. V. Sukumara (✉)
Prapati, No. 36, 2nd Main Road, Kannappa Nagar, Thiruvanmiyur, Chennai, India
e-mail: project.prapati@gmail.com

V. L. Sonavane
Maharashtra Electricity Regulatory Commission, MSEDCL, Mumbai, India
e-mail: vlsonavane@gmail.com

© Springer Nature Singapore Pte Ltd. 2018
R. K. Pillai et al. (eds.), *ISGW 2017: Compendium of Technical Papers*, Lecture Notes in Electrical Engineering 487, https://doi.org/10.1007/978-981-10-8249-8_6

1 Introduction

Convergence of computing, communicating, and control has given lots of comfort to the people in our country. It is used to be a big hassle, even to get encumbrance certificate on the property, but now the same is being delivered to home through Internet. Hon. Prime Minister's program of digitization is bringing Indian rural population to the virtual world. With rise in economy, due to opportunities available with education, even in rural and semi-urban population, the demand for comfort has gone up. Usage of air conditioners, induction heaters for cooking, microwave oven, etc., by rural and semi-urban population have put additional burden on electricity demand from the year 1990 to 2015, and domestic consumption has gone up from 15.16 to 23.53%. Depleting of water table on agricultural feeders has heavy impact on the grid. Long length of wire higher current due to higher demand on agricultural feeder has resulted in higher voltage drop on the lines as well as higher distribution losses.

Sudden surge in the population of urban cities in India needs revamping of existing infrastructure in the available space, which is an additional problem faced by the distribution companies. Single-phase loads on the urban circuits have created unbalance on the lines, and the neutral current has substantially gone up. We are using four core cables instead of three and half core to reduce voltage drop and distribution losses. Peak power demand, current harmonics, overload on the circuits has resulted in poor quality of power distribution. Frequent power interruption, blowing of fuses due to overload, and long time for reconnection have put consumers to lot of discomforts. Power grid is overloaded to the point, and the State Governments have resorted to notifying the public of scheduled power outages, so as to enable consumers to plan accordingly. In fact, the power distribution companies in most states have introduced the concept of a "Power Holiday" a rationing system, whereby industries are asked to take mandatory holidays during different days of the week, so as to reduce the load on the power grid.

In addition to this high volume of equipment in power distribution below, 22/11 kV voltage level is manually operated and requires a lot of effort and struggle to meet consumers' demand. Keeping consumers' interest in mind, some regulators have started imposing penalty for power interruption. When our Honorable Prime Minister stresses on "Make in India" and attracts MNC to make our country as a manufacturing hub, the quality of power supply becomes paramount importance.

There are three primary causes for this situation:

A. *Utilization*: The rapid growth of the economy over the past decade has led to a huge demand for power. Rising demand for power is manifested in major areas: An emerging middle class demands comforts such as washing machines, air conditioners, induction heaters, grinding machines, computers.

The rapid growth of the information technology (IT) sector in India has led to a population exodus toward major hubs, resulting in issues such as sporadic construction of houses and apartments and resulting in substantial demand for power and water supply (which ultimately increases power requirement).

A significant rise in the price of cooking gas has led to people switching to electrical gadgets such as water heaters, electric cookers, induction heaters, and microwave ovens especially in rural areas.

There are rise in the utilization of mass transit systems and battery-operated vehicles within metropolitan areas.

Depletion of water table results in higher power needed for the agriculture.

Rampant power theft in numerous parts of the country.

B. *Government Regulations and environment*: Dependency on supply of coal, failure of monsoon, environmental hazardous for expansion of nuclear power generation, etc., have forced the gap between supplies and demand. GOI is putting all-out efforts to bridge the gap. There is aggressive rise in solar power generation, which is not available during peak demand hours that are from 6 to 10 AM and 6 to 10 PM.

The pace of restructuring of the available distribution infrastructure is slower than the pace, at which the demand is growing.

C. *Subsidization*: This is primarily done in the form of free/subsidized power provided to farmers to run agricultural pumps. This problem has been exacerbated over the past decade due to depletion of water table and use of inefficient motors.

2 Distribution Losses and Quality of Power Distribution

There is a distinct difference in the quality of power in urban and rural (semi-urban) power distribution. In urban power distribution, most of the technical losses are due to unbalance in the circuits, overloading of circuits, the current harmonics, and overloading of distribution transformer. The non availibility of space to enhance the capacity of transformer increases the load and losses.

As regards power distribution in rural and semi-urban areas, covered area is large. Length of wires to be drawn is also high. Almost all states give subsidized or free power to agricultural and rural/semi-urban residential consumers. As a policy, it is proposed to provide single-phase power for 24 h/day and three-phase power for agriculture for 8–10 h. Seeing the success of Jyotigram Yojana, many states have segregated the agricultural feeders to meet the requirement of 24 h for single-phase and 6–8 h for three-phase consumers. HT distribution in rural area runs over long distances, sometimes up to 30 km. When the 11 kV feeders are fully loaded, the line voltage drops. This leads to low voltage on tail end. In turn, this phenomena leads to increase in losses and overheating in motors, electrical gadgets, and transformers, thereby significantly shortening the life span of such devices. Even industrial and commercial feeders which are lengthy have similar problems. The voltage variation on 11 kV line affects many sensitive equipment and has very heavy impact on their performance.

3 Solutions

Distribution automation through "smart power solution" is an answer to many of the problems mentioned above. For HT (22/11 kV) and LT (0.415 kV) power distribution, smart grid and smart power solution are normally limited to smart metering and data acquisition. Because of the absence of cost-effective reliable last mile connections, smart power solution is not in place. Introduction of SCADA/DMS on LT circuits will definitely improve the quality of power and reduce the distribution losses; however, it is a costly proposition. Periodic maintenance of field equipment is a major issue, in which it is very common to bypass the contactor or molded circuit breakers in the distribution transformer outgoing feeder, due to burning of contacts. Basically, the contacts burn due to overload and lack of maintenance.

There is an urgent need for reliable, remote-controlled, maintenance-free, cost-effective switchgear to implement distribution automation using smart power solution through SCADA on LT lines.

Research conducted at IIT Kanpur indicates that by providing switches at different points in the distribution network, it is possible to isolate certain loads, when required. The only option available in the present distribution network is the circuit breaker (one each for every 11 kV feeder) at the 33/11 kV substations. However, these circuit breakers are actually provided as a means of protection to completely isolate the downstream network, in the event of a fault. Using this as a tool for load management is not desirable, as it disconnects the power supply to a very large segment of consumers on the entire 11 kV feeder. Clearly, there is a need to put in place a system which can achieve a more precise load management.

The need for SCADA/DMS at distribution level to improve the quality of power is emphasized by the Executive Director (IPDS) Power Finance Corporation Limited in her letter to all MD's/CMD' of Discoms/Heads Of Power Department vide reference No 02:10:IPDS:2016:Utilities/44476 dated 26DEC2016 and 44552 dated 30DEC2016.

Prapati a "knowledge-based start-up research and development company" has developed a series of spark-/arc-free switches, to convert a "dumb distribution network" to a "smart distribution network."

1. Remote-controlled spark-/arc-free smart grid LT switch useful in urban power distribution,

2. Remote-controlled spark-/arc-free smart grid transposing switch to balance the load on all the lines to reduce current, losses, and voltage drop in neutral conductor,
3. Remote-controlled spark-/arc-free smart grid HT switch to substantially improve tail end voltage and reduce distribution losses to be mounted on existing infrastructure of 11/22 kV load-breaking switch,
4. Remote-controlled spark-/arc-free smart grid LT switch to substantially improve tail end voltage and reduce distribution losses mounted inside the tank of DTR.

Common feature of the switches:

1. Switches are spark-/arc-free switches. There is no spark or arc while opening or closing and no contact bouncing.
2. Switches open during current zerocrossing and close during voltage zero-crossing. No additional harmonics due to switches,
3. Switches are programmable to operate through SMS or power line. The time of closing and opening can be set through SMS or powerline,
4. Switches are single phase and can be operated in single-phase or three-phase modes. All switches are tamperproof (either inside the oil or hermetically sealed),
5. Switches are programmed to operate, on overload with an option to reclose. Additional protection and control can be added,
6. Multiple terminals will eliminate the problem of loose connection, making it totally maintenance free,
7. All retrofit does not need any additional space.

The benefits of a smart grid on LT distribution system are numerous:

Increased use of digital information and control technology, to improve reliability, security, and efficiency of the electric grid.
Dynamic optimization of grid operations and resources with full cybersecurity.
Deployment of smart technologies, appliances, and consumer devices for metering, communications, and distribution automation.
Management of peak load capacity. Improvement in quality of power.
Reduction in distribution losses (**payback period is less than 3 years**)
Enable active participation by consumers.
Anticipate and respond to system disturbances.
Integration of generation, transmission, and distribution systems to enable improved overall grid operations without manual intervention
Improved reliability and "self-healing" of the distribution system.

4 Solution to Various Challenges in Indian Power Distribution

Urban power distribution has the following challenges:

1. Non-availability of space to enhance the capacity of transformer and associated switchgear to meet growing demand and to avoid overload and burning,
2. Unbalance in load current in all the three phases,
3. Increase in neutral current, losses, and voltage drop,
4. Frequent fuse failure due to overload. Replacement to be carried out manually, increasing the downtime,
5. In present arrangement, all the three phases will disconnect during fault increasing the time of power failure,
6. Failure due to loose connections,
7. Wire snap,
8. Neutral wire disconnection,
9. Temporary line-to-line **fault during rainy season.**

5 Solution

5.1 Remote-Controlled Spark-/Arc-Free Smart Grid Switch (SGS)

- SGS does not require additional space to mount. Space is required to mount ACB, and fuses can be saved,
- Since the switch opens in less than 4–5 cycles, with proper coordination, it can be made sure that SGS opens before HV fuse blows, no need to replace HV fuse, under fault condition at the consumer end,
- HV protection can be provided through fuse to protect against internal fault of transformer, and all other LV faults are protected by secondary switch; hence, additional cost of HV breaker and space required to provide HV breaker can be saved,
- Space available can be used to increase the capacity of transformer, to meet the growing demand,
- Heat generated inside the transformer is reduced,
- Each SGS is single-phase unit; hence, during fault, that particular phase will disconnect the load. Downtime will come down to one third,
- Due to auto-reclosure, the point at which the fault has occurred will open after two to three attempts and power will be available to other consumers on the faulty circuit also. Downtime is drastically reduced,

- Hermetically sealed and explosion-proof quality of the SGS makes these switches highly suitable for outdoor duty and hazardous environment,
- SGS can be operated through remote, and data can be collected at remote concentrators,
- Parallel paths will ensure availability of power, even if one terminal loses connection resulting in reduction in downtime,
- Temperature rise on the switch is less than 30° above ambient, making it safe to function even in summer,
- Drop from terminal to terminal is less than 250 mV, and rated current reduces losses inside the switch,
- Even top oil temperature and oil level can be sensed: for preventive maintenance and protection,
- Wire snap, neutral disconnection, earth leakage, higher earth potential can be sensed, and the circuit can be made to trip,

Present installation for 250 kVA transformer:

The basic function of the above can be achieved by pole-mounted smart grid switch (SGS) with less space and cost

Internet of things:

- The smart switch can be operated, remotely using a smartphone.
- It can be connected to a network through the GSM board.
- The network is password protected and can be used by an authorized person only.
- The switch sends data to a centralized monitoring system.
- The centralized system can send commands to switch on real-time basis.
- Each transformer load can be monitored remotely for improved performance/efficiency.
- Through statistical analysis, unauthorized drawing of power can be detected and action can be taken if needed.
- The data will enable to reduce huge cost on inventory and maintenance.

During forced load-shedding, instead of de-energizing entire 11 kV feeder, load on selected DTR can be switched off, so that complete darkness can be avoided to improve safety.

Unbalance in the lines and increase in neutral current can be minimized by introducing "**remote-controlled spark-/arc-free smart grid transposing switch**" to balance the load on all the lines and to reduce current, losses, and voltage drop in neutral conductor.

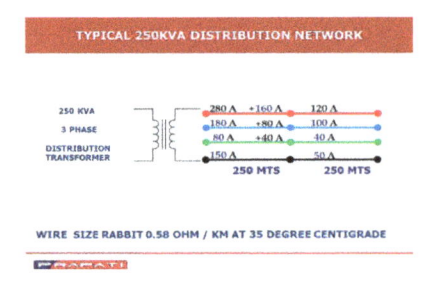

TYPICAL 250KVA DISTRIBUTION NETWORK

250 KVA 3 PHASE DISTRIBUTION TRANSFORMER

280 A	+160 A	120 A
180 A	+80 A	100 A
80 A	+40 A	40 A
150 A		50 A

250 MTS   250 MTS

WIRE SIZE RABBIT 0.58 OHM / KM AT 35 DEGREE CENTIGRADE

Losses before installation of load-balancing unit

R phase = 280 × 280 × 0.145 = 11.36 kW
Y phase = 180 × 180 × 0.145 = 4.7 kW
Phase = 80 × 80 × 0.145 = 0.93 kW
Neutral = 150 × 150 × 0.145 = 3.26 kW
Total = 20.25 kW

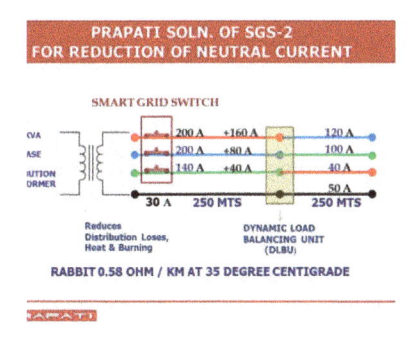

Reduction in distribution losses

R phase = 200 × 200 × 0.145 = 5.8 kW
Y phase = 200 × 200 × 0.145 = 5.8 kW
B phase = 140 × 140 × 0.145 = 2.84 kW
Neutral = 30 × 30 × 0.145 = 0.13 kW
Total losses = 14.57 kW

Total savings/day = 20.25 − 14.5 = 5.75 kW × 10 (Peak hours). Annual saving = 57.5 × 365 = 20,987.5 units

Introduction of rooftop solar power generator with grid-interactive inverter will add more unbalance on the line, and this is a challenge which even many UK power distribution companies are facing.

Load-balancing and reduction in neutral current reduce line voltage drops. Hence, tail end voltage will improve and distribution losses will reduce substantially.

Pole mounted remote controlled MBB transposing switch

Switch is in "make" position before "break"; hence, there is break in power supply to the consumer while transposing the lines.

Problem Faced in Rural Power Distribution

1. Low Voltage at tail end of distribution,
2. High distribution losses,
3. Unauthorized power tapping,
4. Poor consumer metering,
5. Continuous overload and burning of distribution transformer,
6. High inventory and cost of replacement.

Remote-controlled spark-/arc-free smart grid HT switch mounted on the existing load break switch and remote-controlled spark-/arc-free smart grid LT switch mounted inside the tank of distribution transformer:

Time-/zone-based power distribution will improve the tail end voltage and reduce distribution losses.

Metering and data acquisition can be implemented.

Communication can be through GSM/GPRS/WIFI, but control can be achieved through power line, i.e., 11 kV lines.

Overload disconnection with auto-reclosure will protect transformer from abuse and minimizes the burning of transformer and cost of replacement.

Short-circuit protection with auto-reclosure will isolate the faulty feeder, and other consumers continue to get power

Through statistical analysis, unauthorized drawing of power can be monitored and action can be taken if needed.

There is no need to provide special transformer to supply single-phase power for 24 h.

Illustrative Case 1: Switch is on 11 kV line and hence difficult to tamper

Illustrative Case 2: Switch is inside the distribution transformer and hence very difficult to tamper.

Present system

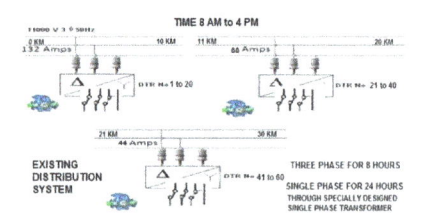

Low-Cost Spark-/Arc-Free Retrofit Smart Grid ...

Proposed time based zone control

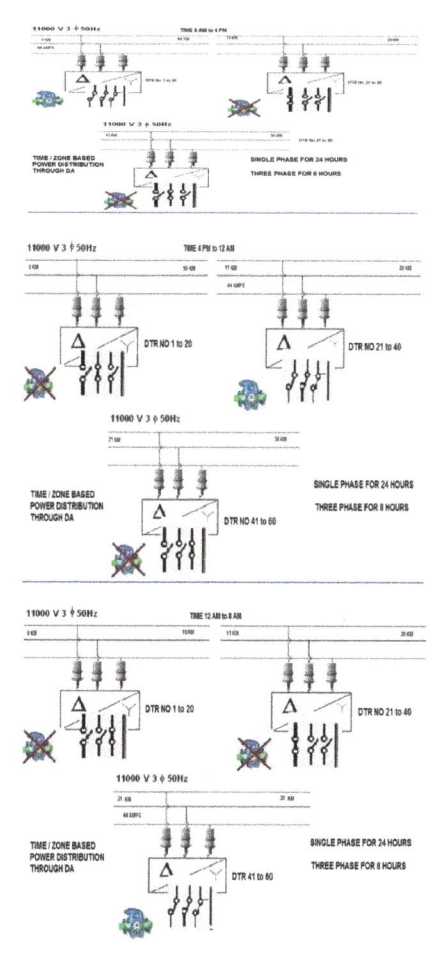

Example: Let make some assumptions:

- Conductor used in AG feeder: 35 mm^2 and resistance of AG feeder conductor = 0.7 Ohms per km,
- If 11 kV feeder length runs for 30 km, and if we divide the zone to 3 section each 10 km, then resistance per 10 km @0.7 Ohms per km = 0.7 × 10 = 7 Ω.
- Designed capacity of 11 kV feeder line = 2500 kVA. Full-load current delivered by each 11 kV feeder = 2500/11 × 1.732 = 131 A,
- Assume the feeder is 100% loaded; then, losses on the wires will be (132 × 132 × 7.0 × 3) = 366 kW + (86 × 86 × 7.0 × 3) = 155 (43 × 43 7.0 × 3) = 38.8 Total losses = 559.8 kW,

- Distribution loss per hour = 559.8 = 560 kWh,
- Energy losses for 8 h per day = 560 × 8 = 4480 kWh per day,
- Losses per annum will be 4480 × 265 = 11,87,200 kWh,
- Voltage drop at tail end: (131 × 7.0) + (86 × 7.0) + (43 × 7.0) = (917 + 602 + 301) = 1820 × 2 (2 line drop) = 3640 V,
- Voltage at tail end = 11,000 − 3640 = 7360 V (7.360 V) on HV and voltage on LV at tail end = 289 V.

Whereas with time- and zone-based distribution:

- The tail end voltage will be 11,000 − (43 × 7.0 × 3) × 2 = (11,000 − 1806) = Volts (9.194 V), voltage at tail end = 361.9 V

Hence, we can conclude that there is a substantial improvement in HT and LT voltages due to time- and zone-based distribution employed in smart grid switches.

6 Conclusions

Many Discoms are comprehensively using **IT** solutions through their outage management system (OMS), which basically uses the **customer information system** (CIS), **geographical information system** (GIS), and **interactive voice response system** (IVRS).

Further, some Discoms have begun to roll out **supervisory control and data acquisition** (SCADA) systems, at distribution level (i.e., at lower voltage levels.)

However, distribution-level SCADA will be effective, only if there is control, and for the control, smart grid switches are absolutely necessary.

Introduction of SCADA/DMS/DA through low-cost spark-/arc-free retrofit smart grid switches will empower distribution managers to introduce e-governance of power distribution.

It will also improve the quality of power and reduce distribution line voltage drops and losses substantially. It will help to manage peak demand, reduce the burning of transformers due to overload, and facilitate instantaneous isolation of faulty circuit.

Micro-phasor Measurement Units (μPMUs) and Its Applications in Smart Distribution Systems

Alok Jain and Suman Bhullar

Abstract The installation of phasor measurement units (PMUs) in the power grids is mainly focused on the transmission systems. Researchers are continuously working on the PMUs for its installation in the transmission system. But, now due to the advancement in distribution system, i.e., for smart distribution system, a device is going to be installed which creates real-time synchrophasor data from the consumer voltage level, called μPMUs, which could provide new insight into modern power systems. These units can be created more cheaply, an order of magnitude less, than current commercial PMUs. For this reason, many more μPMUs could be deployed and provide a much higher resolution of the distribution grid. So, in this paper, the concepts like why we need micro-PMUs and its block diagram are explained. This paper also discusses the applications for synchrophasor technology in distribution systems.

Keywords Discrete Fourier transform (DFT) · Global positioning system (GPS) μPMU · PDCs · WAMS

1 Introduction

A μPMU is a phasor measurement unit that measures time-synchronized voltage and phase angle at high sample rates ∼30/s for transmission and 120/s for distribution. The μPMU is a power quality recording instrument with GPS receiver to enable highly accurate time-stamping for voltage and phase angle measurement. Conventional PMUs in use for the transmission system have ±1° accuracy,

---

A. Jain (✉)
Department of Electrical Engineering, Indian Institute of Technology (BHU), Varanasi, India
e-mail: alok.rs.eee14@itbhu.ac.in

S. Bhullar
Department of Electrical & Instrumentation Engineering, Thapar University, Patiala Punjab, India

© Springer Nature Singapore Pte Ltd. 2018
R. K. Pillai et al. (eds.), *ISGW 2017: Compendium of Technical Papers*, Lecture Notes in Electrical Engineering 487, https://doi.org/10.1007/978-981-10-8249-8_7

and μPMU has 0.01°. A multiphase power flow model and state estimation for distribution systems have been formulated, and solution methods have been presented. It also addresses the following issues like modeling, implementation, observability, and performance [1]. The problem of state estimation in very large power systems has been investigated which may contain several control areas [2]. An approach, which leaves the traditional state estimation software in place, discusses a novel method of incorporating the phasor measurements, and the results of the traditional state estimator in a post-processing linear estimator have been discussed [3]. The benefits of using PMUs for selected real-time applications and present ongoing pilot projects and experience worldwide, and to give short- and long-term road map for future acts have been discussed [4, 5]. Different measurement techniques have been developed by employing global positioning system (GPS) receivers that are suited to the continuous monitoring of the electrical quantities in distribution networks in terms of synchronized phasors [6]. A brief introduction to the PMU and wide-area measurement system (WAMS) technology has been presented and also discusses the uses of these measurements for improved monitoring, protection, and control of power networks [7]. Diverse concepts for the next generation of power distribution system have been summarized whose objective is to bring distribution engineering more closely aligned to smart grid philosophy [8]. Phasor measurement units (PMUs) were installed in an industrial distribution network, and synchronized measurements taken from the PMU at medium and low voltage during different system states were analyzed [9]. Different applications based on energy accounting on the acquisition of the phasor measurements using phasor measurement units (PMU) have been discussed [10]. The developments and present status of the Indian power grid have been discussed and also explore the key areas in which a wider deployment of PMUs may be utilized to make the national grid smart [11]. A new probabilistic approach of the real-time state estimation on the micro-grid has been presented [12]. The performances of a phasor measurement unit (PMU) prototype based on a synchrophasor estimation algorithm conceived for the monitoring of active distribution networks have been introduced [13]. A new technique for estimation of distribution network load model parameters based on PMU measurements data and harmony search algorithm (HSA) have been proposed [14]. The meter placement problem for the measurement infrastructure of an active distribution network, where heterogeneous measurements provided by phasor measurement units (PMUs) and other advanced measurement systems such as smart metering systems are used in addition to measurements that are typical of distribution networks has been studied [15]. The contribution of PMUs that can provide to handling different challenges by examining recent research results on applications of synchrophasors in distribution systems has been examined [16]. The state estimator of all buses in a three-phase network has been presented and proposed greedy algorithm and integer programming optimization method to determine the optimal solution [17]. A mixed-integer programming formulation of DSE that is capable of simultaneously discarding predicted values whenever sudden changes in the system state are detected has been used [18]. Developing and testing frequency-adaptive PMU algorithms with wider linearity range than specified in IEEE Std C37.118-1 have been evaluated in [19], by means of three different concepts, i.e., FIR bandpass filtering, extended Kalman filtering (EKF), and discrete Fourier transform (DFT). The impact of the uncertainties (in terms of phase and magnitude) introduced by arbitrary PMUs on a state estimation process performed on the IEEE 13-bus distribution test feeder has been analyzed [20, 21].

2 Why There Is Need of μPMUS?

Higher degree of accuracy is required for distribution as the angle differences and changes are significantly smaller than in transmission because of the different X/R ratios. μPMU plays an important role in distribution planning and operations as measurement of phase angle and difference in angle between points provides the ability to calculate impedance which is not possible without the PMU. Phase angle also gives information on the direction of power flow for analysis of topology changes or errors. Line level measurement represents an improvement over smart metering for estimating loads on a per phase basis. Today, PMUs are almost exclusively used in high-voltage power transmission. Distribution system applications are more challenging, in three important respects:

1. The voltage angle differences between two locations on a distribution circuit will tend to be at least an order of magnitude smaller than those on the transmission network, because the power flows are much smaller, and the reactance between points of interest is also much smaller. Consequently, meaningful measurement of phase angle differences on distribution systems requires much higher precision—meaning more precise time-stamping and shorter latencies in every step of the transfer of the measurement. Accuracies of the PMUs can still vary by ±1°, while the proposed μPMU technology is expected to discern angle differences to an accuracy of better than ±0.05°.
2. Distribution system measurements will be fraught with much more noise from which the signal must be extracted. This is simply due to the proximity of a large number of different devices connected per mile of circuit at the distribution level, including loads as well as utility switchgear, transformers, capacitors, that may introduce harmonic distortion and transients. Consequently, the background "noise" must be analyzed carefully along with the angle measurements and with the same time stamp. This functionality is not yet standard in PMUs for transmission, where power disturbances are sufficiently infrequent that it is not necessary to explicitly link PMU data to a range of concurrent electrical events. To interpret the rare major disturbance, angle data must then be cross-referenced with data from other monitoring devices after the fact.
3. The economic value of transmission power flows means that larger investments can be justified, with less pressure on the acceptable costs of instrumentation as well as data transmission and concentration. By comparison, the installed costs

must be far lower to make a reasonable business case for the installation of multiple PMUs on a distribution circuit, simply based on the amount of connected load whose service quality or reliability would stand to benefit from increased visibility and better understanding of the dynamics on the circuit.

4. The PMU device can be connected to single- or three-phase secondary distribution circuits up to 690 V (line-to-line) or 400 V (line-to-neutral), either into standard outlets or through potential transformers (PTs) as are already found at distribution substations or could be added on primary distribution circuits if necessary.

3 Micro-phasor Measurement Unit

Micro-phasor measurement unit (μPMU) technology provides phasor information (both magnitude and phase angle) in real time, and the data provided by PMUs are very accurate and enable system analysts to determine the exact sequence of events which have led to the blackouts and malfunctions that may have contributed to the catastrophic failure of the power system. One of the most important features of the PMU technology is that the measurements are time-stamped at the source due to which data transmission speed is no longer a critical parameter in making use of this data. All PMU measurements with the same time stamp are used to infer the state of the power system at the instant defined by the time stamp. It is clear that μPMU data could arrive at a central location at different times depending upon the propagation delays of the communication channel in use. The time tags associated with the phasor data provide an indexing tool which helps create a coherent picture of the power system out of such data.

A. *Phasor and its representation*:

Consider a pure sinusoidal quantity given by

$$x(t) = \sqrt{2} X \operatorname{Sin}(\omega t + \varphi) \tag{1}$$

ω being the frequency of the signal in radian per second, and φ being the phase angle in radians.

Equation (1) can also be written as

$$x(t) = \operatorname{Re}\{X e^{j(\omega t + \varphi)}\} = \operatorname{Re}[\{e^{j(\omega t)}\} X e^{j\varphi}] \tag{2}$$

The sinusoid of Eq. (1) is represented by a complex number X^* known as its phasor representation

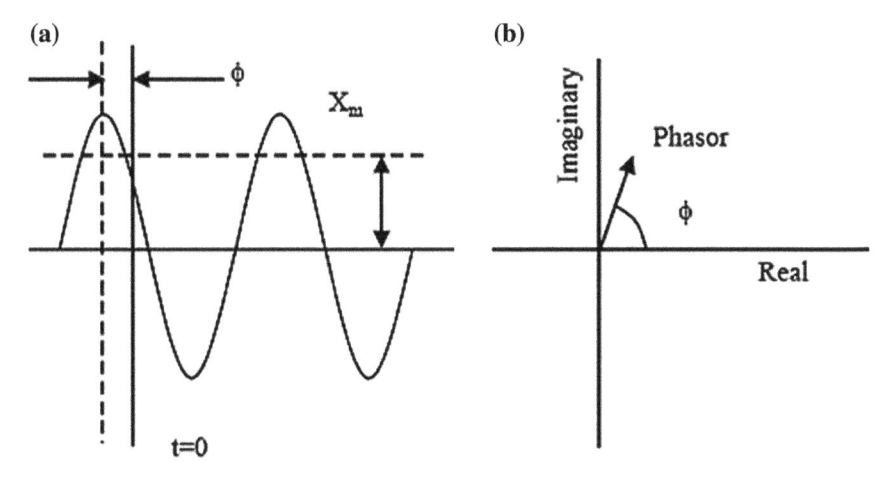

Fig. 1 **a** A sinusoid. **b** Its phasor representation

$$X^* = Xe^{j\varphi} = X[\cos\varphi + j\sin\varphi] \quad (3)$$

A sinusoid and its phasor representation are shown in Fig. 1a, b. The phase angle of the phasor is arbitrary, as it depends upon the choice of the axis $t = 0$. Note that the length of the phasor is equal to the RMS value of the sinusoid.

B. *Block Diagram of Micro-phasor Measurement Unit*:

Phasor measurement unit (PMU) technology provides phasor information (both magnitude and phase angle) in real time, and its block diagram is shown in Fig. 2. The analog inputs are currents and voltages obtained from the secondary windings of the current and voltage transformers. The current and voltage signals are converted to voltages with appropriate shunts or instrument transformers so that they are matched with the requirements of the analog-to-digital converters. The sampling

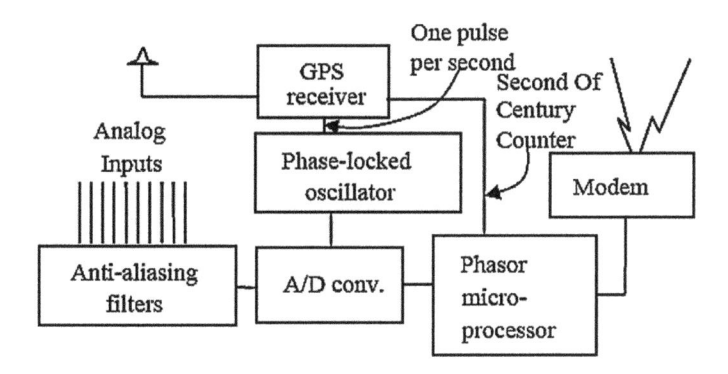

Fig. 2 Block diagram of the μPMU

rate chosen for the sampling process dictates the frequency response of the anti-aliasing filters. In most cases, these are analog-type filters with a cutoff frequency less than half the sampling frequency in order to satisfy the Nyquist criterion. The sampling clock is phase-locked with the GPS clock pulse. Sampling rate can be used 12 samples per cycle of the nominal power frequency to as high as 96 or 128 samples per cycle. The microprocessor calculates positive-sequence estimates of all the current and voltage signals.

C. *Global Positioning System*:

The most important use of the GPS system is to generate the signal of one pulse-per-second. This pulse when received by any receiver on earth is coincident with all other received pulses within 1 µs. In practice, much better accuracies of synchronization of the order of a few hundred nanoseconds have been realized. There are four satellites in each of the six orbital planes displaced from each other by 60° and having an inclination of about 55° with respect to the equatorial plane, which orbit around the earth with a period of half a day as shown in Fig. 3. The GPS satellites keep accurate clocks which provide the one pulse-per-second signal.

D. *Hierarchy for PMUs*:

The PMUs are situated in power system substations and provide measurements of time-stamped positive-sequence voltages and currents as well as frequency and rate of change of frequency of all monitored buses and feeders. The measurements are stored in local data storage devices, which can be accessed from remote locations for diagnostic purposes. The phasor data are also available for real-time applications. At the next level, phasor data concentrators (PDCs) are present as shown in Fig. 4, which gather and record the data, reject altered data, and align the time stamps from several PMUs. PDCs have storage facilities and application functions which need the PMU data available at the PDC. This can be made available by the PDCs to the local applications in real time. An another level of the

Fig. 3 GPS satellite

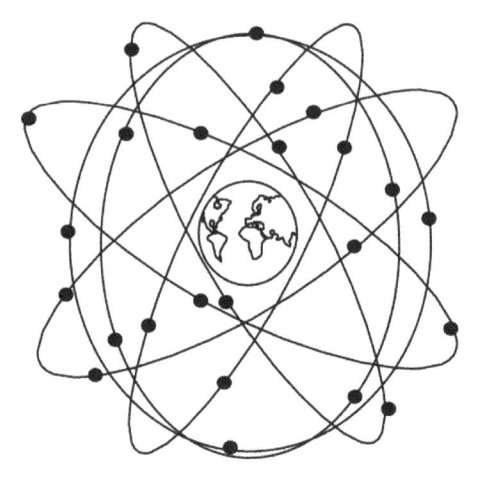

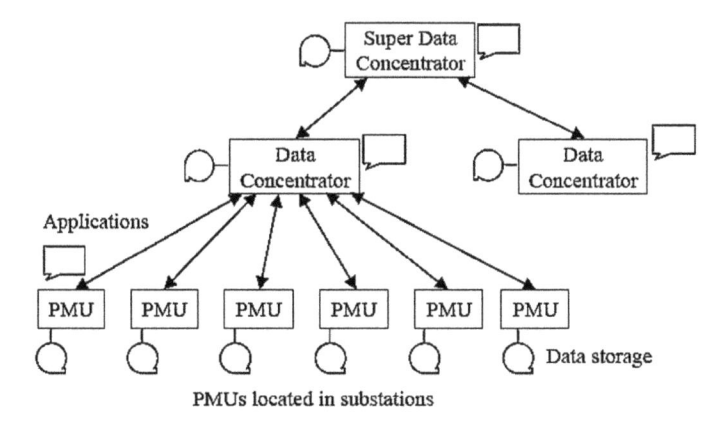

Fig. 4 Hierarchy of the PMUs and PDCs

hierarchy is called super data concentrator (SDC) where there is facility for data storage of data aligned with time tags as well as a steady stream of near real-time data for applications which require data over the entire system. Figure 4 shows the communication links to be bidirectional. As, most of the data flow is upward in the hierarchy, although there are some tasks which require communication capability in the reverse direction.

E. *Communication links for PMUs*:

Communication facilities are essential for applications requiring phasor data at remote locations. Generally, two types of data transfer are used in any communication task. Channel capacity is the measure of the data rate that can be sustained on the available data link. The second aspect is the latency, defined as the time lag between the time at which the data is created and when it is available for the desired application. Diagnostic analysis applications require PMU data to help in analyzing the power system performance during major disturbances. Leased telephone circuits were among the first communication media used for these purposes. Switched telephone circuits can be used when data transfer latency is not of importance. Electric utility communication media like power line carrier and microwave links have also been used. But, now the fiber-optic links is used as the medium which have high data transfer rates, unsurpassed channel capacity, and immunity to electromagnetic interference.

4 Distribution Applications

Diagnostic applications and control applications are broadly classified applications of micro-PMUs, and it is useful to distinguish diagnostic from control applications using μPMU data to help operators better understand the present or past condition of the distribution system, or to inform specific control actions to be taken (likely by automated systems) in more or less "real time."

A. *Diagnostic Applications*:

Diagnostic applications for consideration includes island detection, fault location and high-impedance fault detection, identification of fault-induced delayed voltage recovery (FIDVR), distribution system state estimation including reverse power flow detection and phase balancing, renewable generation monitoring, oscillation detection, characterization of generator inertia, supporting transmission system diagnostics, and many more.

Some of them are explained below:

(1) *Island Detection*:

Today's inverters have very reliable anti-islanding protection. However, with greater penetration of diverse distributed resources and more complex dynamics on distribution circuits, it may become increasingly difficult-to-distinguish fault events from other abnormal conditions where it is desirable to keep DG online (e.g., low-voltage ride-through). The comparison of phase angle between a potential island and the rest of the grid is the most definitive test that offers not only high sensitivity, but specificity—i.e., ruling out an island if the phase angle remains locked, and thus allowing generators to remain online when they are needed most. Preventing DG from unnecessary trips during stressed grid conditions has important implications not only for distribution power quality and reliability but for transmission operators as well, who are increasingly concerned about the vulnerability of the grid to cascading events behind the substation.

(2) *Fault Detection and Location*:

Protective devices on distribution circuits are generally based on overcurrent relays that respond to a combination of current magnitude and duration. This makes it very difficult to detect high-impedance faults, where the fault current is similar in magnitude to load current. Furthermore, once a fault is isolated, its exact location is difficult to determine remotely.

The actuation of a particular circuit breaker or fuse only identifies a general section of a feeder where the fault has occurred. The standard approach then is for line crews to physically patrol the length of the faulted line section, looking for damaged equipment. This process is time-consuming and costly, even more so for underground cables.

Algorithms exist for recognizing high-impedance faults as well as for locating faults through proper analysis of monitored data, but the quality of available measurements on distribution circuits is often insufficient to support them. We expect that μPMUs will allow fault detection and location with much greater precision than before, even with relatively few devices deployed on a circuit. This is because voltage angle measurement makes it possible to compute changes in impedance between two measured points and thus diagnose a fault even if the current magnitude is insufficient to trip a protective relay. The impedance between the faulted point and a PMU on either side then also indicates the relative location of the fault. If successful, methods based on μPMU measurements could drastically reduce service restoration times and enhance safety by ensuring reliable fault detection.

(3) *Fault-Induced Delayed Voltage Recovery (FIDVR)*:

It is a condition marked by a prolonged period of voltage recovery after a low-voltage event due to a relatively brief fault, followed by voltage overshoot and a period of high voltage. FIDVR is caused primarily by single-phase residential air conditioners that stall under low voltage and draw large currents before triggering a thermal switch that trips them off.

The resulting loss of load then causes a high voltage condition, which can in turn trigger corrective devices such as switched capacitor banks and create further instability. FIDVR is a problem identified at the transmission level, but will likely be resolved in distribution where the problem originates. The diagnostic challenge, for which μPMU data may be suitable, is to quickly distinguish the unique characteristics of FIDVR from other types of abnormal voltage conditions, or even anticipate a FIDVR event, so as to avoid overcorrection and perhaps develop effective active mitigation measures.

(4) *Postmortem Analysis*:

In case of the need for analysis after a fault, the recorded synchrophasor measurements are very valuable. With these "flight recorder data" available, a quick evaluation of the facts can be performed. Intelligent user interfaces enable the operator to quickly find the cause and sequences of the disturbance. Also the documentation of the event, for example, on request of authorities is possible in a fast and easy way. This application is valuable also for use in distribution networks.

(5) *Voltage Stability Monitoring*:

Voltage stability monitoring is one of the standard applications for synchrophasor measurements. In transmission systems, this application monitors the load of a transmission line or corridor, using PMU measurements on both ends. In distribution systems, there are normally no explicit transmission corridors. However, monitoring of the dynamics of the voltages gives a good picture of the reactive power flow.

(6) *State Estimation*:

State estimation, or identifying the steady-state voltage magnitude and phase angle at each node in a network, significantly informs the situational awareness of human operators as well as many automated control actions in a power system. However, state estimation is generally more difficult for distribution than for transmission systems. This is because distribution systems are harder to model (owing to untransposed lines with phase imbalances, small X/R ratios, large numbers of connecting load points, and less redundancy from Kirchhoff's laws) and present a high-dimensional mathematical problem, while at the same time offering few physical measurements to inform the state estimation. Direct voltage angle measurements on a feeder could vastly speed up and improve the accuracy of state estimation techniques.

(7) *Reverse Power Flow*:

A simple yet important aspect of the distribution system operating state is reverse power flow on any line segment. The significance of reverse flow hinges on the type of protection system design used by the utility, and whether the coordination of protective devices could be compromised under reverse flow conditions. While some circuits may be able to safely backfeed all the way through the substation, others could introduce problems that would be expensive to remedy with bidirectional protection. One way to take advantage of μPMU data would be to detect reverse power flow on any feeder segment with a minimal placement of physical devices throughout the circuit.

Owing to the information conveyed by phase angle, fewer points may have to be instrumented than with conventional current measurement, potentially making the μPMU approach more economical.

(8) *Renewable Generation Monitoring*:

Besides line flows, a key aspect of situational awareness for distribution operators is knowledge of generation resources and loads on a circuit. With much DG connected behind the meter, however, only *net* loads are visible to the operator. This type of masking of generation and load compromises forecasting for both sides of the equation and makes it difficult if not impossible to estimate N−1 contingencies (such as a common-mode generation trip following a disturbance). It is conceivable that analysis of μPMU data could help "unmask" net metered generation to assist in both distribution system operation and planning.

B. *Control Applications*:

Control applications include protective relaying under two-way flow, volt-VAR optimization, coordination of resources on a micro-grid, intentional islanding and resynchronization of micro-grids, and the creative recruitment of distributed resources for ancillary services.

(1) *Protective Relaying*:

To employ different protection schemes that accommodate reverse flow safely is one of the techniques. Developing the develop supervisory differential relaying schemes based on μPMU data that does not require a costly replacement of protective devices.

(2) *Volt-VAR Optimization*:

We do not expect that voltage angle measurement would afford an inherent advantage over magnitude for feeder voltage optimization, but the capability to support this important function alongside other applications could add significantly to the business case for μPnet deployment.

(3) *Micro-grid Coordination*:

To advance the opportunities for active control based on μPMU measurements, we will study requirements for hierarchical, layered, distributed control of an islandable cluster of aggregated distributed resources and identify the merits, if any, of angle as a state variable. Micro-grid balancing and synchronization is an application with a longer strategic time horizon, but one where the use of voltage angle as a control variable is expected to be crucial.

5 Conclusions

These micro-phasor measurement units can be created more cheaply, an order of magnitude less, than current commercial PMUs. For this reason, many more μPMUs could be deployed and provide a much higher resolution of the distribution grid. So, in this paper, the concepts like why we need micro-PMUs and its block diagram are explained with the applications for synchrophasor technology in distribution systems.

References

1. Meliopoulos APS, Zhang F (1996) Multiphase power flow and state estimation for power distribution systems. IEEE Trans Power Sys 11(2):939–946
2. Zhao L, Abur A (2005) Multiarea state estimation using synchronized phasor measurements. IEEE Trans Power Syst 20(2):456–462
3. Zhou M, Centeno VA, Thorp JS, Phadke AG (2006) An alternative for including phasor measurements in state estimators. IEEE Trans Power Syst 21(4):1930–1937
4. Skok S, Ivankovic I, Cerina Z (2007) Applications based on PMU technology for improved power system utilization. In: IEEE power engineering society general meeting
5. Tate JE, Overbye TJ (2008) Line outage detection using phasor angle measurements. IEEE Trans Power Syst 23(4):1644–1652
6. Carta A, Locci N, Muscas C, Sulis S (2008) A flexible GPS-based system for synchronized phasor measurement in electric distribution networks. IEEE Trans Instrum Meas 57(11): 2450–2456
7. Carta A, Locci N, Muscas C (2009) A PMU for the measurement of synchronized harmonic phasors in three-phase distribution networks. IEEE Trans Instrum Meas 58(10):3723–3730
8. De La Ree J, Centeno V, Thorp JS, Phadke AG (2010) Synchronized phasor measurement applications in power systems. IEEE Trans Smart Grid 1(1):20–27
9. Heydt GT (2010) The next generation of power distribution systems. IEEE Trans Smart Grid 1(3):225–235
10. Naumann A, Komarnicki P, Powalko M, Styczynski ZA, Blumschein J, Kereit M (2010) Experience with PMUs in industrial distribution networks. In: Power and energy society general meeting. IEEE
11. Wache M, Murray DC (2011) Application of synchrophasor measurements for distribution networks. In: Power and energy society general meeting. IEEE

12. Rihan M, Ahmad M, Beg MS (2011) Phasor measurement units in the indian smart grid. In: Innovative smart grid technologies India (ISGT India). IEEE PES
13. Chenine M, Vanfretti L, Bengtsson S, Nordström L (2011) Implementation of an experimental wide-area monitoring platform for development of synchronized phasor measurement applications. In: Power and energy society general meeting. IEEE
14. Ding F, Booth CD (2011) Applications of PMUs in power distribution networks with distributed generation. In: Proceedings of the 46th international universities' power engineering conference (UPEC)
15. Bahabadi HB, Mirzaei A, Moallem M (2011) Optimal placement of phasor measurement units for harmonic state estimation in unbalanced distribution system using genetic algorithms. In: 21st International conference on systems engineering (ICSEng)
16. Alinejad B, Akbari M, Kazemi H (2012) PMU-Based distribution network load modeling using harmony search algorithm. In: Proceedings of 17th conference on electrical power distribution networks (EPDC)
17. Ding F, Booth CD (2012) Protection and stability assessment in future distribution networks using PMUs. In: 11th International conference on developments in power systems protection, DPSP
18. Sánchez-Ayala G, Agüero JR, Elizondo D, Lelic M (2013) Current trends on applications of PMUs in distribution systems. In: Proceedings of innovative smart grid technologies. IEEE, Washington, DC, pp 1–6
19. Vipin Krishna R, Ashok S, Krishnan MG (2014) Synchronized phasor measurement unit. In: Proceedings of IEEE international conference on power, signals, controls and computation, Kerala, pp 1–6
20. Aminifar F, Shahidehpour M, Fotuhi-Firuzabad M, Kamalinia S (2014) Power system dynamic state estimation with synchronized phasor measurements. IEEE Trans Instrum Meas 63(2):352–363
21. Colangelo D, Zanni L, Pignati M, Romano P (2015) Architecture and characterization of a calibrator for PMUs operating in power distribution systems. In: IEEE International conference on PowerTech, Eindhoven, pp 1–6

A Rule-Driven Architecture to Address Interoperability in an IEC 61850 Series-Based Power Utility Automation System

Mayank Sharma and Thomas Rudolph

Abstract The IEC 61850 series of standard has become the choice of standard to build power utility automation systems. With the use of standard file format exchange during engineering process, definition of information models, services and mapping them over a standard communication interface, the vision of a truly multi-vendor automation solution has been made possible. Further, the stated goal of IEC 61850 series is to reach interoperability between functions to be performed by power utility automation in a multi-vendor environment. In addition to the standardized engineering process using file exchanges, the use of analytic on the top can bring additional benefits in terms of increasing system specification efficiency, improve tendering phase of project and identify and mitigate risks such as detecting non-interoperable behaviour at an early stage of project life cycle. In this paper, an enhancement around the use of rule-driven engineering approach is proposed. It is demonstrated how the use of analytic using present generation vendor agnostic engineering tools could be used to detect non-interoperable behaviour for a proposed automation system. Further, a cloud-based approach to engineering offers data consistency and collaborative framework over entire system life cycle.

Keywords IEC · 61850 · Smart grid · Utility automation system
Substation automation system · Interoperability · Performance · Profile
BAP · SGAM

M. Sharma (✉)
Innovation & Architecture—Energy Automation, Schneider Electric, Noida, India
e-mail: mayank.sharma@schneider-electric.com

T. Rudolph
Energy Automation, Schneider Electric, Frankfurt, Germany
e-mail: thomas.rudolph@schneider-electric.com

© Springer Nature Singapore Pte Ltd. 2018
R. K. Pillai et al. (eds.), *ISGW 2017: Compendium of Technical Papers*, Lecture Notes in Electrical Engineering 487, https://doi.org/10.1007/978-981-10-8249-8_8

1 Introduction

The power system design and operation has undergone a profound change in past decade under several emerging trends. Distributed energy resources are being integrated at all voltage levels leading to an increased requirement for better systems and software to mange power flow in the grid. Government policies mandate required level of renewable addition every year. Technical advancements have allowed a near monotonic reduction in component pricing and optimized cost of implementing a given technology. Business models have evolved taking into account the smart end consumer and provides monetary incentives on smart utilization of energy.

Energy is no more unidirectional in traditional sense but has become bidirectional with complex participation of different stakeholders in the energy business. This business calls for a different approach in the way such systems are designed and maintained. At the same time, the system design should be able to capture the business process and application cases in order to extract the information exchange requirement that will allow a system and system of systems to interoperate and be secure to evolutions.

IEC 61850 series of standard [1] has become a standard of choice for designing power utility automation system. The main value of IEC 61850 is not just the mapping to a communication stack; it is the detailed support of business-related semantic models which allow modelling of power system equipment, associated "smart" functions and related global information exchange. Furthermore, it provides enough flexibility to address new demands as demonstrated by the recent evolution towards the smart grid, as depicted in the smart grid architectural model (SGAM), adopted by IEC 62559 and IEC 62913-1 [2, 3]. The SGAM clearly specifies the outreach of the IEC 61850 to all domains concerning power and energy sector and all the way up to the control centre zone. The interfacing with zones covering operational, enterprise and market dynamics is modelled by the IEC 61968 and IEC 61970 [4, 5] (Fig. 1).

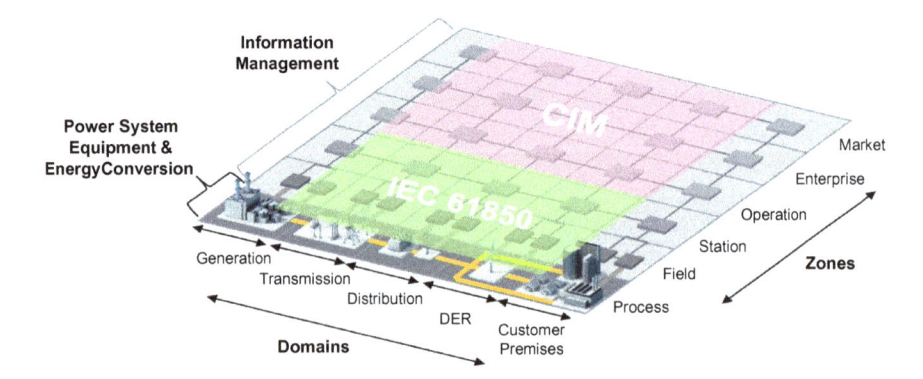

Fig. 1 SGAM layer

2 System Engineering and Lifecycle Aspects

IEC 61850-6 defines the complete system engineering life cycle—from system specification, to pre-engineering IEDs, to automation system engineering. A system specification begins with a description of the system—an electrical one-line diagram, and allocation of functional elements (logical nodes) to the parts and equipment associated to the one line. Then, the IEDs provide a defined set of logical nodes that are bound to a specific process function and primary equipment. Finally, service models are selected to address the information sharing functions and are mapped to a communication protocol. Now the elements are in place to build an automation system.

A system configuration description language (SCL) for configuration of power utility IEDs is provided. One of the key purposes of the SCL is to ensure an interoperable exchange of communication system configuration data between IED engineering tools and a system engineering tool in a multi-vendor environment. The SCL is based on Extensible Mark-up Language (XML) and consists of several files, e.g. system specification description (SSD), system configuration description (SCD), system exchange description (SED), IED capability description (ICD) and configured IED description (CID). These are used during the configuration of IEC 61850-based power utility automation systems as shown in Fig. 2.

As the IEC 61850 standard has evolved, so have engineering tools. First-generation tools provided some open interfaces to integrate third-party IEDs. Later, manufacturer-independent tools followed the IEC 61850 standard. Tools like Schneider Electric's System Engineering Toolsuite (SET) represent a third-generation of these tools as they facilitate system specification and integration by exchanging engineering information via standardized file formats on a single PC.

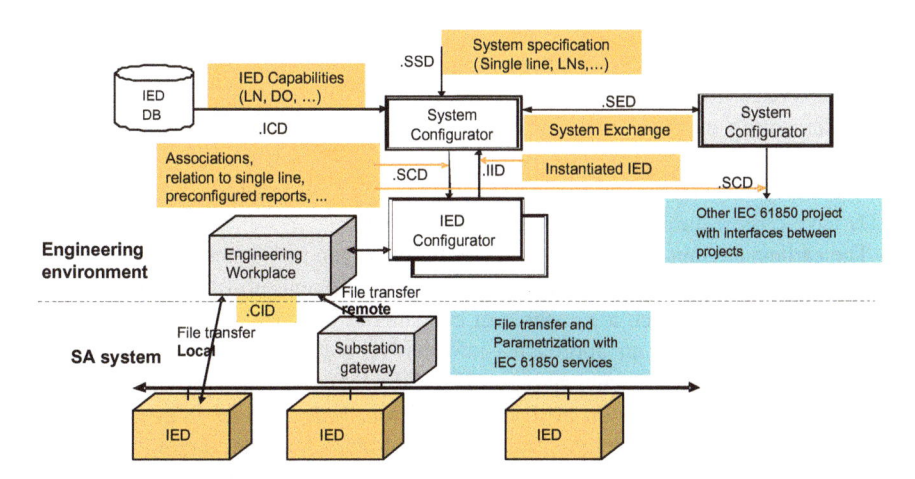

Fig. 2 IEC 61850 series-based engineering process

Most of these tools focus on the pure integration aspect. The next generation of tools will provide further assistance by helping users to reduce capital expenditures (CAPEX) and operational expenditures (OPEX) of system integration. These benefits are achieved by adding business rules which provide engineering knowledge, by introducing new analytics that reduce technical risks during system specification and design phase. New requirements such as user (group) specific profiling of standard elements can be configured as new business rules are applied. Further use of cloud based architecture enable collaborative experience by knowledge sharing amongst multiple stakeholders acorss several geographical locations.

3 Interoperability Aspects

Interoperability according to the IEC 61850 series of standard is defined from the aspect of being syntactically interoperable (devices connected to a common bus and common protocol), semantically interoperable (a device understanding information provided from other devices) and lastly free allocation of functions in devices to achieve distributed functionality. Further, the approach to interoperability, clause 6.5 of the IEC 61850 part 5, edition 2.0 states—"the requested mutual understanding of devices from different suppliers result in proper data and communication service model (IEC 61850-7-x). The mapping of this model to state-of-the-art communication stack (coding/decoding) shall be defined unambiguously in IEC 61850-8-x and IEC 61850-9-x series. It should be noted that interoperability is not a device property but a system goal".

Despite free allocation of functions and standard syntax and semantics, interoperability still remains a major challenge for system integrators and utility owners operating in a multi-vendor environment [6]. From a system engineering perspective, the IEC 61850 standard provides a high degree of flexibility in terms of data model implementation hosted within an intelligent electronic device (IED). This means that each IED has a different information support and service model in terms of implementing optional fields/features within the IEC 61850 standard. Thus, a subset of IEC 61850 can reside in different products and solutions, which potentially could lead to interoperability issues when using multi-vendor IEDs even if the IEDs have passed the IEC 61850 conformance test.

The earliest attempt to address interoperability is seen in the standard itself. IEC 61850-5 defines application functions comprising of different sub-functions called as logical nodes (LNs) which are capable of exchanging information with each other. The Annex F provides an approach—informative but not binding—to function description. Later, a draft technical report IEC TR 61850-7-500 [7] provides a set of guidelines and explanation on how to use different logical nodes defined in IEC 61850-7-4 and model applications for substation domain. Another ongoing work of TC 57 WG 10 outlines a concept of basic application profile (BAP) which is based on domain-specific basic application functions (IEC TR 61850-7-500). These profiles are created based upon agreed-upon selection and interpretation of relevant parts of the IEC 61850 standard with the intention of building interoperable user/project specification [8]. BAPs include application description (from IEC 61850-5, IEC TR 61850-7-500), communication services and BAP-specific PICS, MICS and PIXIT.

A harmonization attempt has been effectuated by IEC TC 57 WG 19 to ensure interoperability in the long term. The upcoming IEC 62361-103 [9] is on standard profiling as applicable to the common information model (CIM) and the IEC 61850 series of standard. This part describes the framework and methodologies to derive profiling concepts from the requirements coming out from specific business process and application functions domains. In the profiling concept for the IEC 61850 series of standard, BAPs will be the standard basic level profiles in the sense of a standard. The framework and methodologies will be valid as well for the project-specific application profiles. Figure 3 depicts the general framework of deriving profiles. A project context uses one or more existing profiles to realize the description of its business process via resulting set of application functions. A profile is derived from a standard, various optional extensions relevant to a group of users and a set of existing profiles with an iterative process of improvement.

Additionally, the profiles follow a set of rules which play an important part during the overall end-to-end project lifecycle process including aspects of specification, engineering, commissioning and maintenance. Such set of rules could create a knowledge base inside an organization, leveraging past project experiences and challenges, collating non-standardized information on device capabilities and translating them into a rule-based formalism and deploying an IT infrastructure hosting such rules and making them available in a collaborative fashion across all stakeholders participating in a given project—tendering, engineering and design, commissioning, etc. Such usage of rules will allow maximization of project specifications and early detection of possible non-interoperable behaviour from communication, semantics and syntactical aspects.

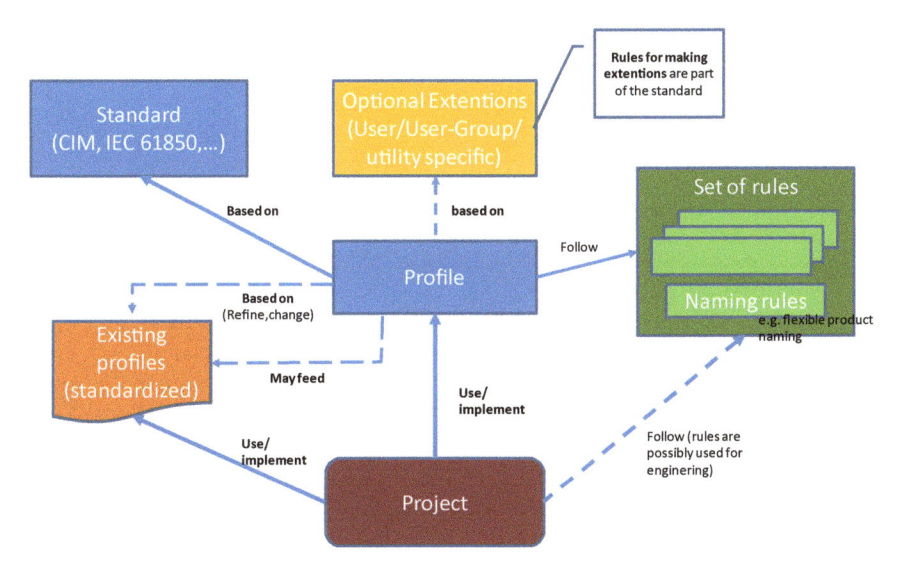

Fig. 3 Framework for profile [9]

4 Rules-Based Analytics for Interoperability on System Level

As described in the previous section, the creation of a set of rule definitions that supplement SCL file exchanges provides an additional way to leverage IED capabilities during system specification. Rules will incorporate aspects on device capabilities and utilizing past knowledge on device integration during past project execution. Such manipulations allow system specification to become more robust because a knowledge base of IEC standard compliant capabilities is accessed along with previous project and systems integration experience. It should be mentioned here that while current open point of a preferred machine-readable format for representing profiles and rules are currently under discussion as part of standardization efforts, this should not act as a discussion to utilize rules to improve efficiency of overall project execution cycle.

An engineering environment provided by tools such as SET could support such extensions proposed by Fig. 3 by proposing a content management system that could host a set of rules supporting the system specification and engineering workflow. Figure 4 proposes such an arrangement where rule creation, storage and access by an engineering interface could be collaborative thanks to a cloud-based hosting environment.

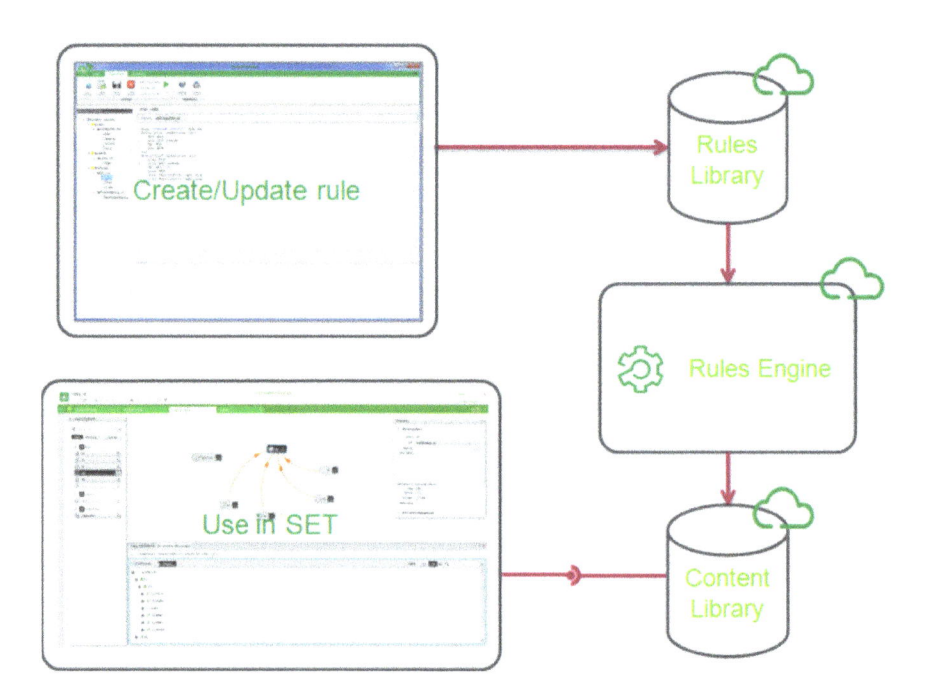

Fig. 4 Architectural context for rules exploitation during engineering

Such rule definitions allow system integrators to create rules in a specific project context and store them for exploitation during the project design phase. This helps during the tendering system design phase for greenfield or retrofit solutions. A system integrator role for solution engineering could be a vendor, a user or a third-party engineering company. In cases where the system integrator role is executed by a user or a third-party engineering company, the rule editing context can be easily shared and applied to an agreed-upon business model. Dynamic rules specific to a given project may also be created and pushed into the rules repository in real time, thus allowing for greater flexibility during project design activities.

Rules can be of myriad types. Starting from a simple rule—e.g. of what optional fields are supported and what are not for a given report control block—going up to rules that may involve comparing several rules to generate a more complex one could be envisaged using a choice of language. Here are some rules with their possible impact:

- A rule based upon the constraints on an IED "A" with respect to the string length of a subscribed GOOSE control block name published from an IED "B". If the string length exceeds a certain threshold, IED "A" cannot subscribe to a GOOSE message from IED "B". This results in a non-interoperable behaviour. A possible representation of such a rule could be as follows:

  Given \<Manufacturer\> \<IED A\> \<Version\>
  Define MaxChar GSEControl.name = 10//for SCL//

- A rule based upon the capabilities on an IED to support the control model as stipulated in the BAP (eg: select-before-operate with enhanced security is enumerated by number 4, where the numbers 0,1,2,3 represent other control models as stipulated in the IEC 61850 series.).

  Given \<Manufacturer\> \<IED A\> \<Version\>
  Define ctlModel = 4

- If an IED does not support the data quality "test", it cannot be used in test mode which could impact the entire test strategy.

  Given \<Manufacturer\> \<IED A\> \<Version\>
  Define LLN0.Beh = 3 && LLN0.Beh = 4

A set of rules that have been selected captures the additional constraints a device or a group of device(s) may have to follow in order for a successful integration with a proposed solution. Figure 5 depicts the availability of those capabilities available inside a given project context.

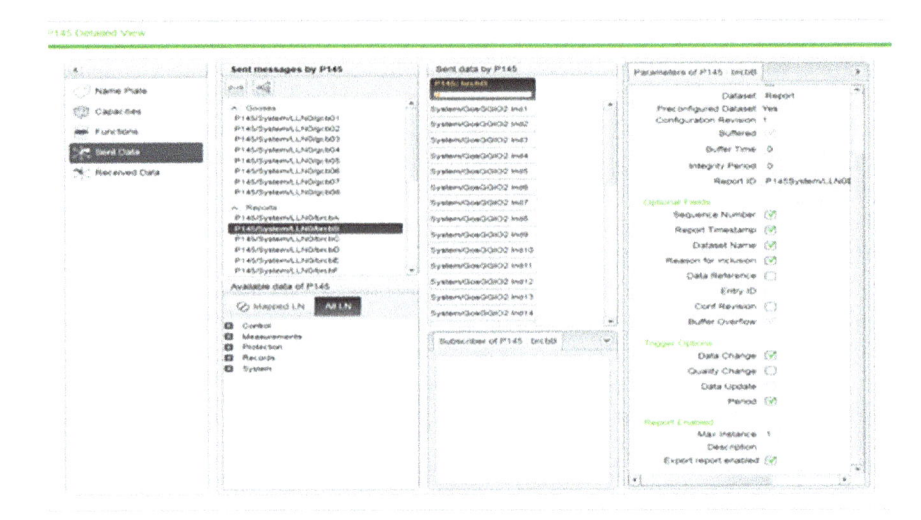

Fig. 5 Rules exploitation inside an engineering tool

5 Conclusion

While standardization bodies have just started this work defining framework and methodologies around profiling and that further work is necessary in this direction, it has been demonstrated in this paper that how engineering tools like SET are supporting the path towards profiling by already providing the capability to apply rules inside a project environment.

References

1. IEC 61850 (2015) Communication networks and systems for power utility automation—ALL PARTS, IEC 61850:2015 SER
2. IEC 62559 (2015) Use case methodology—ALL PARTS, IEC 62559:2015 SER
3. IEC 62913-1 (2013) Generic smart grid requirements—specific application of the use case methodology for defining generic smart grid requirements according to the IEC system approach, IEC 62913-1
4. IEC 61968 (2015) Application integration at electric utilities—system interfaces for distribution management—ALL PARTS, IEC 61968
5. IEC 61970 (2015) Energy management system application program interface (EMS-API)—ALL PARTS, IEC 61970
6. Holbach J, Rodriguez J, Wester C, Baigent D, Frisk L, Kunsman S, Hossenlopp L (2007) Status on the first IEC 61850 based protection and control, multi-vendor project in the United States, Power systems conference: advanced metering, protection, control, communication, and distributed resources, 2007. PSC 2007, pp 254, 277, 13–16 March 2007

7. IEC TR 61850-7-500 (2016) Use of logical nodes for modeling applications and related concepts and guidelines for substations, IEC DTR 61850-7-500
8. Guise L, Huon G, Lhuiller P, Haecker M, Brunner C (2014) IEC 61850 interoperability at information level. A challenge for all market players, CIGRE 2014 session, Paris
9. IEC 62361-103 (2016) Power system management and associated information exchange—standard profiling, IEC CDV 62361–103

Software Defined Networking for Smart Grid Communications and Security Challenges

M. U. Shaileshwari, K. S. Nandini Prasad and A. Paventhan

Abstract With the aim of empowering smarter energy usage and integration of renewable distributed energy resources (DERs), smart grid has been proposed as an evolution of the current power systems leveraging the most advanced information and communication technologies (ICTs) to provide an intelligent bi-directional electricity and communication network. Smart grid is a large-scale, heterogeneous, and distributed network, which poses many challenges to be overcome from communication networking to autonomous control and management. In recent years, the paradigm of software defined networking (SDN) has attracted much attention. It proposes a new concept of networking architecture which abstracts the control functionalities from the packet forwarding hardware (data plane) to an external software controller (control plane). This is extremely convenient for large data centers to cope with virtual machine networking in which virtual machines are created dynamically and move between different physical machines. Due to the controller being implemented as software and its programmatic interfaces to individual networking devices are exposed to other software applications, any network applications and services based on such an architecture can be more agile. Furthermore, application systems are enabled to be network-aware, which means that they are aware of the properties, requirements, and state of the network environment and can quickly adapt to changes in the network context. Therefore, SDN is perceived to have tremendous potential for utilization by the underlying communication infrastructure of the smart grid. With the advent of SDN, the interface between applications and networks will be greatly changed. Applications

---

M. U. Shaileshwari (✉)
Utility Automation Research Centre, Central Power Research Institute (CPRI),
Bangalore, India
e-mail: shaileshwari@cpri.in

K. S. Nandini Prasad
Department of ISE, Dr. Ambedkar Institute of Technology, Bangalore, India
e-mail: nandiniks1@gmail.com

A. Paventhan
Centre Head (Bangalore & Chennai), ERNET India, Bangalore, India
e-mail: paventhan@eis.ernet.in

in smart grid will have a higher degree of network awareness which enables more dynamic interactions with the underlying network. In this paper, the challenges and security concerns are presented to explore the opportunities for SDN technology in smart grid communication.

Keywords Smart grid · Software defined networking · Cyber security Cyber-physical system

1 Introduction

The Indian Power Sector is undergoing an enormous changes in terms of information and communication technology (ICT). With regard to energy efficiency, another buzzword often conjured is that of smart—smart meters, smart grids, smart cities [1].

The challenges that these smart systems for managing energy consumption is the risk they pose to the data that is collected, stored, and managed by smart systems—both in terms of security and privacy [2]. Therefore, in order to guarantee safe and stable operation, it becomes essential to closely monitor the power system and intervene more frequently to countervail imbalances. Further, increasing numbers of intelligent electronic devices (IEDs) will have to be installed in the grid, which enable processing, exchanging monitoring, and control data. This leads to a larger data volumes and a high degree of complexity which needs to be handled by the ICT infrastructure [3]. Thus, the successful growth toward smart grid requires high-performance communication networks, which enable reliable, robust, time critical, and secure data transmission.

The underlying infrastructure of smart grid must be effective and reliable in transmitting large amounts of real-time data, scalable and flexible in aggregating resources, and secure and convenient in providing management interfaces to upper layer application systems [4].

Therefore, new rapidly growing SDN technology is perceived to have tremendous potential for utilization by the underlying communication infrastructure of the smart grid.

1.1 Concepts of Software Defined Networking (SDN)

Software defined networking (SDN) is a network architecture where the control plane is decoupled from the hardware and implemented as a software application. Legacy routers and switches had both the control plane and the data forwarding plane implemented in the same hardware appliance. SDN architecture separates the

Fig. 1 Architecture of a legacy router/switch

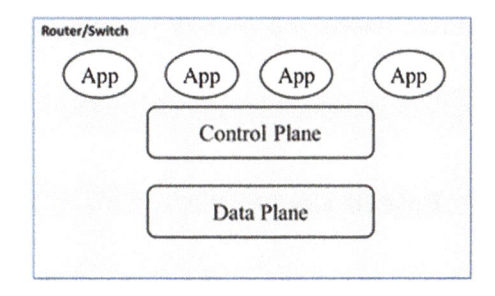

Fig. 2 Typical architecture of an SDN

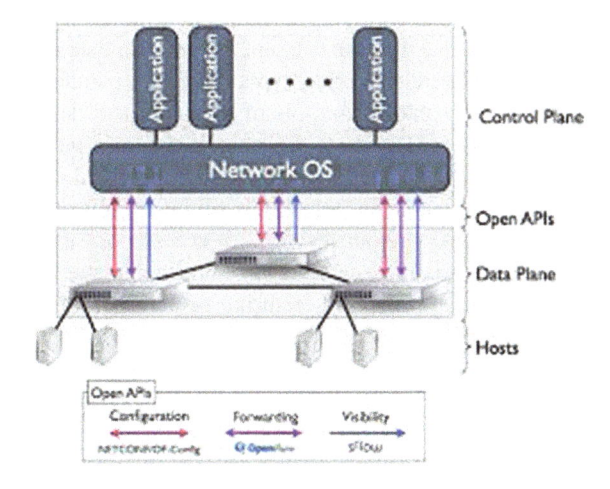

two and makes the control plane run on any standard server in a centralized location. This architecture provides more programmability and control to the network administrators without requiring physical access to the network's hardware devices that are involved in data delivery.

SDN has transformed the networks from following a tightly coupled architecture to a distributed architecture. Figures 1 and 2 describe the architectural differentiation between legacy networks and software defined networks.

1.2 OpenFlow

SDN requires some protocols for the control plane to communicate with the data plane. One of such protocols is OpenFlow [5]. OpenFlow is the first standard communication interface defined between the control and forwarding layers of an SDN architecture. It allows direct access to and manipulation of the forwarding plane of network devices such as switches and routers, both physical and virtual. In other words, it is a communication protocol proposed for supplying communication between routers and network switches (data plane) to routing decision center (control plane). It simply gives programmer control over routing protocol of routers [5–7].

2 Ease of Use

The rapid increase of smart devices with their ability to access the Internet created new opportunities to build large-scale cyber-physical systems. This, however, poses new challenges for not only keeping up with the dynamicity of the hardware and software, but also with the resilient and reliable data collection in such systems. The emerging software defined networking (SDN) paradigm can perfectly address these challenges by splitting controls of networks and data flow operations. One of the major goals in SDN is to be able to interact with the networking equipment (e.g., routers, switches) to create an open networking architecture for everyone. In this way, one can get a global view of the entire network and will be able to make global changes without having to access to each device's unique hardware. Eventually, various large-scale network architectures can be deployed and maintained with ease while still featuring resiliency and robustness (Fig. 3).

The emerging SDN paradigm can provide excellent opportunity for reducing the network management cost by integrating a software-based control that can be flexible with respect to software upgrades, flow control, security patching, and quality of service (QoS).

SDNs have potential to offer greater security by providing consistent access control, ability to apply security policies efficiently and effectively, and the ability to centrally manage and control network topology.

Fig. 3 OpenFlow protocol stands

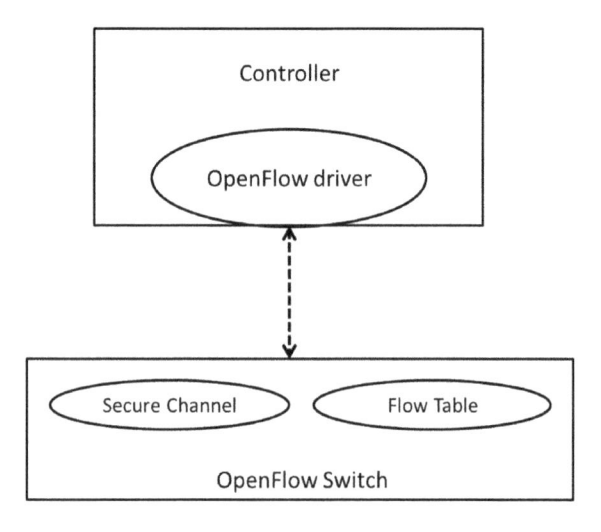

3 Need for SDN-Enabled Smart Grid Communication

The smart grid communication infrastructure mainly uses network components that are owned and operated by utilities and energy companies. In some cases, they also lease services from telecommunication companies or third-party cloud services. In any case, the management of the networks is a great challenge due to the scale. Furthermore, due to using different vendors and applications, the equipment's may not be interoperable. Therefore, the utilities will need to deal with equipment maintenance and software upgrades that bring a lot of burden in terms of cost and labor. The following reasons necessitate need for a more flexible network management technology based on SDN:

- In a large-scale network, software upgrades are challenging.
- Seamless integration of multiple networks.
- New equipment and improved technology necessitate replacing the software without any disruption to the existing older equipment.
- And many more network communication challenges.

SDN provides solutions to these problems due to the following advantages it brings [8]:

- SDN provides global view of the network which makes the large-scale management more effective.
- SDN adopts open standards and introduces technology abstraction, which provides a vendor-agnostic approach to configuring and maintaining various types of network elements that are common in smart grid.
- Hardware virtualization through SDN eases the burden of managing different networks while using resources efficiently.
- Due to its holistic view of network, the SDN-based network will provide superior control of delay and jitter in the network which is crucial for SCADA systems in terms of state estimation and control.

4 Opportunities for SDN in Smart Grid Communication

Smart grid has a large-scale communication network that consists of home area networks (HANs), neighborhood area networks (NANs), and wide area networks (WANs). Each of these network components deploys thousands of devices such as intelligent electronic devices, phasor measurement units (PMUs), routers, switches, and computers [8]. In addition, these networks employ different underlying technologies whether it be wired or wireless along with various communication protocols. This brings up the problems of interoperability as well as network management.

Software defined networking (SDN) provides very increasing features which will revolutionize the future networks like centralize control mechanism, cost efficiency, innovation, programmability, scalability, security, virtualization, cloud support, and efficient environment to support big data [9].

The advances in SDN technology, the existing power grid is going through a massive transformation to make it smarter (i.e., smart grid), which will be more reliable and connected with the ability to transfer data and power in two ways. The need for data communication in the power grid necessitated upgrading the existing grid network infrastructure with different components such as home area networks (HANs), neighborhood area networks (NANs), and wide area networks (WANs). Each of these smart grid applications deploys thousands of network devices that need to be managed continuously for reliable operations. Unfortunately, the management of this massive infrastructure requires additional labor and cost for the utility companies who manage these networks.

One sustainable solution to this network management problem is the use of the emerging SDN, which can provide excellent opportunities for reducing the network management cost by integrating a software-based control that can be flexible with respect to software upgrades, flow control, security patching, and quality of service (QoS) [10]. Nonetheless, while a significant amount of work has been done in the SDN space, most of these efforts targeted the applications in the area of cloud computing, data centers, and virtualization and there is a need to adapt SDN for the existing needs of smart grid applications.

In addition, the data communication motivation necessitates upgrading the existing network infrastructure with different components. With these new transformations, smart grid systems will need to maintain a large-scale heterogeneous network that brings a number of challenges. One of the challenges is the ability of this networking infrastructure to self-heal itself during man-made or natural (e.g., hurricane, earthquake) disasters so that potential blackouts and temporary outages can be minimized through continuous interactions between different components of the smart grid, the new energy infrastructure should reconfigure the control of the physical assets and network topology in an efficient manner and achieve resilient operations. Each of these networks deploys thousands of network devices that need to be managed continuously. Unfortunately, this massive infrastructure requires additional labor and cost for the utility companies who own and manage these networks. The emerging SDN paradigm can provide excellent opportunity for reducing the network management cost by integrating a software-based control that can be flexible with respect to software upgrades, flow control, security patching, and quality of service (QoS).

4.1 Use Case

Before moving into the description of how SDN can be used for smart grid applications, we first briefly explain the existing smart grid network infrastructure.

Basically, there are three major components in a smart grid infrastructure. (1) Home area networks (HANs) that mainly connect home devices with the smart meters; (2) neighborhood area networks (NANs) which collect smart meter data from houses; and (3) wide area networks (WANs) which provide long-haul communication with the utility control centers using various technologies including cellular ones.

Smart grid control center will collect data through communication network from consumers, field devices such as PLCs, PMUs, IEDs at substations in real time and do control decision at the control center in terms of reliability and quality of the power. A control center consists of number of servers each defined for performing its functionality, networking components, LAN, WAN, routers, switches, firewalls, etc., as shown in Fig. 4.

In traditional system, to provision data center for some new application configuration updates which requires to be updated on all servers as shown in Fig. 4.

In case the user needs to apply configuration settings to all the servers deployed in the data center, all user's needs to access the server through telnetting each server one after the other and configure the necessary settings. Obviously, there is a lot of overhead here and there may be chances of misconfiguring at least one of the servers.

There is a better way to automatically push the configurations to the systems that we would need in order to provision these new applications. So, this problem with traditional configuration of servers in the data center is the heart of SDN. To make it easier and to make it more consistent in order to enact changes in the network environment that we need.

The basis for SDN is the technology of breaking the control plane of operations from the data plane in a router as shown below in Fig. 5.

The router functionality can be separated in the following manner. The control plane is an IoS operating system like working in conjunction with CPU. Now the data plane is the forwarding intelligence for the device, and this is implemented typically in application-specific integrated circuits (ASICs) which forwards or rejects the packets [5].

By doing the separation, we have a decision-making process or control of the device and separate this from forwarding. When we do this, we can control security on control plane in a separate and distinct manner how we would control data plane

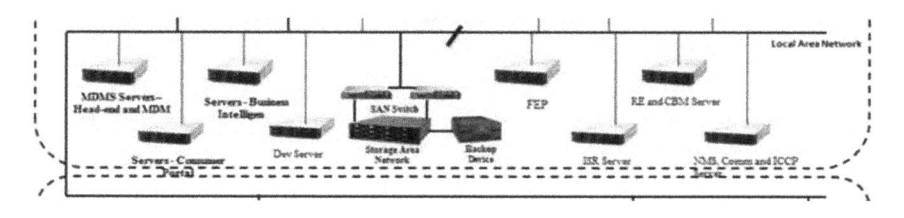

Fig. 4 Conceptual deployment of servers in smart grid data center

Fig. 5 Separating router functionality to control and data plane

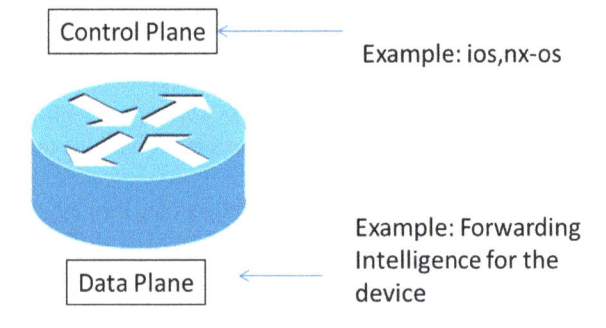

Fig. 6 SDN model (separating the control plane of each of devices)

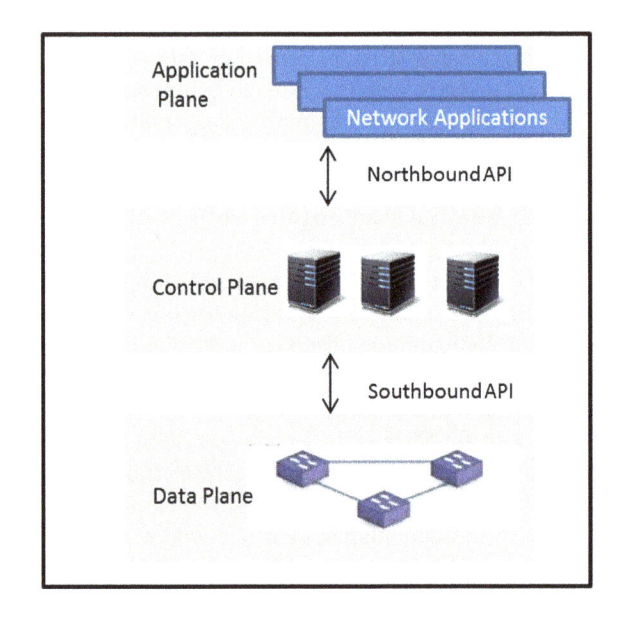

traffic. We can optimize control plane functions in a separate and distinct manner how we would optimize data plane functions.

To make SDN a reality, we would take the control plane from all over many devices that we make up a network infrastructure and we can control them centrally from a SDN controller component as shown in Fig. 6. When we do this, the application is to be configured to all servers/devices. Now, all the applications will make a request for configuration changes of the SDN controller, it will in turn reach to all of the control plane of the many devices that will make the data center, and it will make sure that they are acting in accordance with the needs of this application.

So, separating all the control plane from the devices and gathering up in the controlling of all those many control planes from one centralized location is a key aspect of SDN.

To communicate with SDN controller and control plane of many devices, we use OpenFlow protocol. This protocol will provide a secure means to communicate from SDN controller to control plane devices.

Since the goal of SDN is to be open and programmable, smart grid requires a specific type of network behavior. Due to this flexibility, we can develop and install an application to behave in a specific type. These applications may be traffic engineering, security, QoS, routing, switching, virtualization, monitoring, load balancing, etc.

When a packet arrives to SDN controller or network operating system, the forwarding device will parse the packet and know what to do with the packet or it will query the SDN controller. SDN controller will decide what action to take on the packet and push this information to the forwarding device.

Advantages: The SDN controller is logically centralized network operating system, the controller has a global view of all network forwarding devices below it and has a way to communicate with packet forwarding instructions through them. The SDN controller can directly be an abstraction or simplified view of network-to-network applications to make key decisions on how to implement network policies.

5 Security Challenges in SDN for Smart Grid Communications

SDNs have potential to offer greater security by providing consistent access control, ability to apply security policies efficiently and effectively, and the ability to centrally manage and control network topology. However, SDN networks introduce some new security vulnerabilities on their own. These vulnerabilities are related to single point of failure in the SDN controller, potential to cause congestion in communications between data and control places.

The threats arising from the security compromise can be passive or active. In passive method, the attacker only monitors (or eavesdrops), records the communication data occurring in the SDN-enabled smart grid, and analyzes the collected data to gain meaningful information. In the active one, the attacker tries to send fake authentication messages, malformed packets, or replay a past communication to the components of the SDN-enabled smart grid. As passive threats are surreptitious, it is harder to catch their existence. However, it is easier to catch the existence of an active attacker, but its damage to the smart grid can be relatively higher than the passive threats.

6 Conclusion

In this paper, we have demonstrated how SDN has been applied for communications in smart grid considering a specific requirement of updating the application configurations in the data center simultaneously to all connected devices. We have also discussed that how the emerging SDN paradigm could be considered as a viable technology for the smart grid communication architecture, which is currently under massive modernization effort by the utility providers. And we have also discussed the security challenges in SDN for SDN-enabled smart grid.

Acknowledgements Authors gratefully acknowledge the Central Power Research Institute, Bangalore, Dr. Ambedkar Institute of Technology, Bangalore, and ERNET India, Bangalore, for technical support provided by them.

References

1. Bonneau M. ICT's role in sustainability, the case of smart grids: potential and challenges
2. Guidelines for Smart Grid cyber security, vol 2. Privacy and the Smart Grid, NIST, August 2010
3. Dorsch N, Kurtz F, Georg H, H¨agerling C, Wietfeld C. Software-defined networking for smart grid communications: applications, challenges and advantages. TU Dortmund University, Communication Networks Institute (CNI), Germany
4. Zhang J, Seet BC, Lie TT. Opportunities for software-defined networking in smart grid. Department of Electrical and Electronic Engineering, Auckland University of Technology, Auckland, New Zealand
5. Open Networking Foundation. https://www.opennetworking.org/
6. Open Daylight. http://www.opendaylight.org
7. Open vSwitch. http://openvswitch.org
8. Aydeger A. Software defined networking for smart grid communications. Florida Interation University, Florida
9. Alam F, Katib I, Alzahrani AS. New networking era: software defined networking. Int J Adv Res Comput Sci Softw Eng. Computer Science Department, Faculty of Computing & IT Kind Abdulaziz University Jeddah, Saudi Arabia
10. Cahn A, Hoyos J, Hulse M, Keller E. Software-defined energy communication networks: from substation automation to future smart grids. University of Colorado, Boulder. In: IEEE SmartGridComm 2013 symposium—smart grid services and management models
11. Sequeira A. Report by Fundamentals of SDN

Digital Utility

Deepak Chaudhary

Abstract Smart grid introduction, challenges faced by the distribution utilities and application of smart grid technologies in TPC-Mumbai.

Keywords Challenges faced by the distribution utilities · Application of smart grid technologies in TPC, Mumbai

1 Introduction

1.1 Smart Grid

A smart grid is an electrical grid which includes a variety of operational and energy measures including smart meters, smart appliances, renewable energy resources, and energy efficiency resources.

1.2 Challenges and the Way Ahead

There are various challenges faced by the distribution utilities in Mumbai, and those can be resolved by using the smart grid solutions.

1. **Energy settlement**: Energy settlement between the Discoms in Mumbai is done on cumulative basis on the last day of every month. The actual readings for the consumers are taken on different days of the month; however, the settlement is done on the last day of every month. To arrive at the settlement units, the units are prorated at the end of every month, which always gives some variation in the settlement units between Discoms. So with the help of smart meters, the

---

D. Chaudhary (✉)
DSS-BILLING, Tata Power Company Limited, Mumbai, India
e-mail: deepak.chaudhary@tatapower.com

accurate reading of all the meters at the end of every month is obtained and chances of any discrepancy in settlement of units every month become negligible.

Also, this activity is carried out during the entire month because of a lot of manual intervention in the calculation of settlement units which adds to the administrative cost.

2. **Load forecasting**: Currently, the real-time load data is not available and load forecasting is done on the basis of past five year trends in day ahead scheduling. With the help of smart meters, real-time data is available and hence load forecasting will be more accurate. This will help in fine tuning power purchase strategy and peak load management.

2 Ease of Use

2.1 Application of Smart Grid Technologies in TPC, Mumbai

Tata power, Mumbai, has taken various initiatives on smart grid technologies, which are produced below.

1. Data from meter installed at T<>D and G<>T boundary is captured on real-time basis in a central server and used for energy accounting and settlement.
2. All consumers are mapped to DT & Feeder and accurate energy. Audit is conducted through SAP by obtaining meter readings through AMR.
3. AMR technology is used for meter reading and billing for high revenue base consumers.
4. Meter data is analyzed using meter data analytics system to identify the faulty meters and unauthorized use of electricity.
5. GIS technology is being used for improved meter reading sequence, bill dispatch sequence, and customer service. Meter reading efficiency has gone up by more than 10% with the meter reading, while a person is able to take more number of meter readings in a single day. Also, improved bill dispatch sequence is helped in improving the number of daily bill delivery and reduced dependency of bill dispatch vender. Whenever there is a fault in consumer vicinity, a complaint is logged for the same, and the front-end executive gets live information on his screen about the fault and he can provide the expected resolution time of the fault. This helps in responding to the consumer's query and increased consumer satisfaction.
6. With the implementation of the smart grid solution, the variation in energy settlement has come down to less than 0.03%.
7. With the implementation of the smart grid solution, the peak load variation is below 5%.

Analysis of Communication Channel Attacks on Control Systems—SCADA in Power Sector

Rajesh Kalluri, Lagineni Mahendra, R. K. Senthil Kumar, G. L. Ganga Prasad and B. S. Bindhumadhava

Abstract Usage of open standard protocols such as IEC 60870-5-101/104 in supervisory control and data acquisition (SCADA) systems which are not provided with security features leaves vulnerabilities for the attacker. Adopting these protocols in power SCADA draws more attention to attacker since a successful attack at one location may lead to catastrophic failure and may lead to a blackout. The critical communication channel in power SCADA is between remote terminal unit (RTU) and master terminal unit (MTU), and they communicate over IEC 60870-5-101/104 protocols. Any successful attack on this channel may lead to disastrous effect [1–3]. Simulation of attacks targeting this communication channel provides a better perception of impact for any successful attack. In this paper, attacks on communication channel have been discussed using an influence diagram. Experiments are also conducted to study the impact of communication channel attacks on power system which is simulated using real-time digital simulator (RTDS). This paper brings out the details of the experiment conducted and the results thus obtained. This paper also discusses countermeasures to protect systems from such kind of attacks.

R. Kalluri (✉) · L. Mahendra · R. K. Senthil Kumar · G. L. Ganga Prasad · B. S. Bindhumadhava
Real Time Systems and Smart Grid Group, Centre for Development of Advanced Computing, Bengaluru, India
e-mail: rajeshk@cdac.in

L. Mahendra
e-mail: laginenim@cdac.in

R. K. Senthil Kumar
e-mail: senthil@cdac.in

G. L. Ganga Prasad
e-mail: gpr@cdac.in

B. S. Bindhumadhava
e-mail: bindhu@cdac.in

Keywords Remote terminal unit (RTU) · Master terminal unit (MTU)
Real-time digital simulator (RTDS) · IEC 60870-5-104/101 protocol and vulnerabilities · Cyber attack · Supervisory control and data acquisition (SCADA)

1 Introduction

Supervisory control and data acquisition (SCADA) systems are used to control dispersed assets using centralized data acquisition and supervisory control. SCADA systems are widely used in critical infrastructure industries like power, water, oil, chemical etc. Traditionally, SCADA systems were running on proprietary hardware and software and were isolated. With advancements in technology of IT systems for better communication, SCADA systems are connected for real-time data exchange over different communication mediums such as very small aperture terminal (VSAT), microwave, wireless local area network (WLAN). Adapting to the rapid growth in technology, SCADA networks are potentially vulnerable to manipulation of critical operational data that could lead to serious disturbances.

As SCADA systems become part of the architecture, providing security procedures is the bigger challenge without affecting the existing architecture. Communication infrastructure may have security issues which are unique to vendor hardware and software as well as protocols used for communication.

Typically, SCADA systems are composed of three components such as RTU, MTU, and communication links. Field devices are connected to RTU over RS-232 and RS-485. RTU acquires data from field devices and acts as a data concentrator. MTU communicates with the RTU over the communication links adhering to the protocol agreed upon by RTU and MTU. Each MTU may communicate with multiple RTUs based on the requirement. Human–machine interface (HMI) is used for data visualization.

MTU and RTU communicate adhering to a protocol. Protocols can be proprietary or open standard. Interoperability is a critical issue with proprietary protocols. Some of the open standard protocols are IEC 60870-5-101 [4], IEC 60870-5-104 [5], DNP 3, Modbus, etc. Adhering to the protocols, MTU can acquire data from RTU for field monitoring as well as can issue commands for needful control.

To protect SCADA systems, it is required to analyze security risks as well as develop security measures. However, it is impractical to conduct an attack in real time. This requires a simulation bench with simulation tools. All components of SCADA including communication channel need to be modeled. But, to understand the transient behavior of electric networks, it is required to have a simulator such as real-time digital simulator (RTDS) [6, 7].

The rest of the paper is organized as follows: Sect. 2 discusses the communication channel attacks on control system, and Sect. 3 discusses representing communication channel attacks using an influence diagram. Section 4 discusses the experiment setup and the components required for analysis of attacks. Section 5 outlines the steady-state simulation. Section 6 discusses two transient simulation cases with the experiment results. Section 7 discusses the countermeasures that can be applied to protect from such types of attacks. Section 8 concludes the paper summarizing the findings of the results.

2 Communication Channel Attacks on Control Systems

Most critical area for communication channel attacks in SCADA environment is communication between MTU and RTU, wherein the IEC 60870-5-101 and IEC 60870-5-104 protocols are used for data acquisition and control. This is one of the major areas of communication of data wherein the modification of data may lead to wrong control decisions which will cause chaos. But, IEC 0870-5-101 and IEC 60870-5-104 protocols are plaintext protocols and are vulnerable to various communication channel-related attacks such as man in the middle, replay, sniffing.

2.1 Difficulties in Simulation of Communication Channel Attacks

Communication protocols used in SCADA systems are not same as IT systems. In case of IT systems, tools such as Ettercap and packet creator are useful to sniff and replay. But in case of SCADA systems, even though an attacker can sniff the data but to tamper/conduct a replay attack, same tools may not be sufficient. Adding to that, an attacker should have the domain knowledge to conduct an attack. The attacker should know about details of protocols, packet structures and how to tamper the data packet at a precise position. Hence, an attacker without knowing the domain knowledge may not be able to conduct an attack on the communication channel of SCADA where RTU and MTU communicate using standard/proprietary protocols.

2.2 Coordinated Attacks

Consider a case where an attacker having knowledge about IT systems such as gaining network access, authentication credentials, protocol specifications, type of communication protocols, details about protocols and not having any details about field such as deployment of sensors. In this case, attacker can sniff the data and can conduct man-in-the-middle attacks using techniques such as address resolution protocol (ARP) poisoning. Conducting an attack on the field without knowing the field details is generally considered as a dumb attack. In case of dumb attack, attacker modifies the payload of a packet without knowing the impact on the field. This scenario leads to an attack on the field, but the chances of a considerable effect on field are very less.

To conduct an attack successfully with a huge impact on field, attacker should know about the field details. Here, field details may include details of sensors, control points, layout diagram, critical points of operation, etc. The attacker may use approaches such as sabotage or phishing to get these details.

Coordinated attacks are generally conducted by a group of people with different skills and domain knowledge. In this scenario, an attacker with communication channel tampering skills and an internal employee with field details cooperatively conduct an attack. When a group of attackers attacks a particular industry/nation for causing damage is generally termed as targeted coordinated attack. In general, targeted coordinated attack is conducted by a group of attackers located in different geographical areas.

3 Influence Diagram for Communication Channel Attacks

Influence diagrams are similar to decision trees, attack trees, or defense graphs [8] and used to represent a problem. These diagrams are useful in analyzing a scenario and useful for planning countermeasures. In this scenario, various scenarios of conducting attacks on communication channel have been represented. A successful attack on communication channel may lead to tampering of data between RTU and MTU, data loss, and commands manipulation which is initiated from MTU to RTU. Some of the critical possible attacks which can be initiated between RTU and MTU are replay, data modification, eavesdropping, masquerade, control bypassing, IP spoofing, session hijacking. Possible attacks have been modeled using influence diagram [9] and are as shown in Fig. 1. Influence diagrams are useful for understanding the attack methodology.

Influence diagrams are useful for planning and representing countermeasures at the appropriate nodes. Providing authentication and encrypting data are some of the countermeasures represented as shown in Fig. 1. Countermeasures are represented in white eclipses. Influence diagrams are also used to represent impact when an attack is successful. The impact is represented in a diamond shape. Loss of data, delay of data, command manipulation, and data tampering are some of the impacts that can be observed when an attack on a communication channel is successful.

To understand influence diagram better, it is required to traverse from bottom to top and follow all the gray blocks. For data modification attack between RTU & MTU, attacker has to follow these steps:

Analysis of Communication Channel ... 119

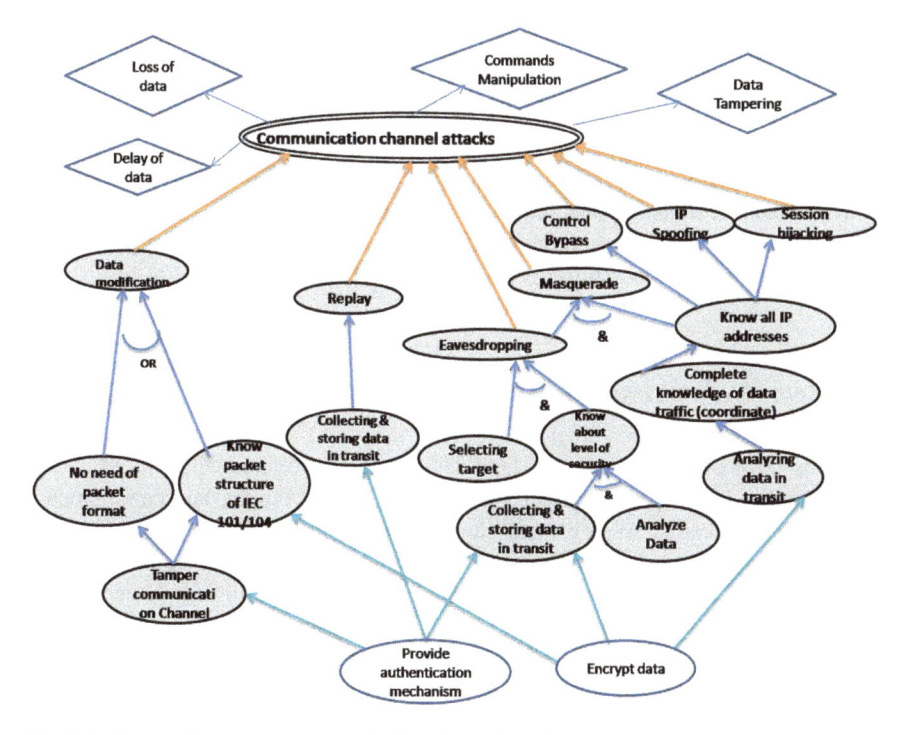

Fig. 1 Influence diagram for communication channel attacks

- Tamper communication channel between RTU and MTU. Since IEC 60870-5-104 protocol is plaintext, all data is visible to attacker.
- Choose any one path, i.e., no need of packet format or know the packet structure of IEC 104. Even though attacker does not know the packet format, data can be tampered randomly or by popping HMI. When an attacker knows about packet format, it is easier to locate critical data and tamper.
- Data modification is successful in any of the selected paths.

To conduct replay attack between RTU and MTU, attacker has to follow these steps

- Tamper communication channel between RTU and MTU. Since IEC 60870-5-104 protocol is plaintext, all data is visible to attacker.
- Since IEC 870-5-104 is plaintext protocol, it is possible to collect and store the data in transit.
- Replay attack will be successful when attacker retransmits same data.

4 Experimental Setup

4.1 Real Time Digital Simulator (RTDS)

RTDS allows accurate and reliable simulations of three-phase electromagnetic and electromechanical transient phenomenon in electric networks both for closed-loop equipment testing and offline simulation studies [6]. RTDS used for simulating the 400-kV Salem substation using RSCAD software (Fig. 2).

4.2 Remote Terminal Unit

RTU is used for acquiring data from field devices. RTU provided with various cards such as analog input, analog output, digital input, digital output, and communication unit (CMU) cards. Analog input card is used for acquiring data from analog devices, and analog output card is used for sending commands to control analog devices. Digital input card is used for acquiring data from digital devices, and digital output card is used for sending commands to control digital devices. CMU cards are used for aggregation of all input data. This data will be communicated to the master station as per a standard protocol. CMU card is also used to receive a command from MTU adhering to the protocol and issue to the respective analog/digital device through the output cards.

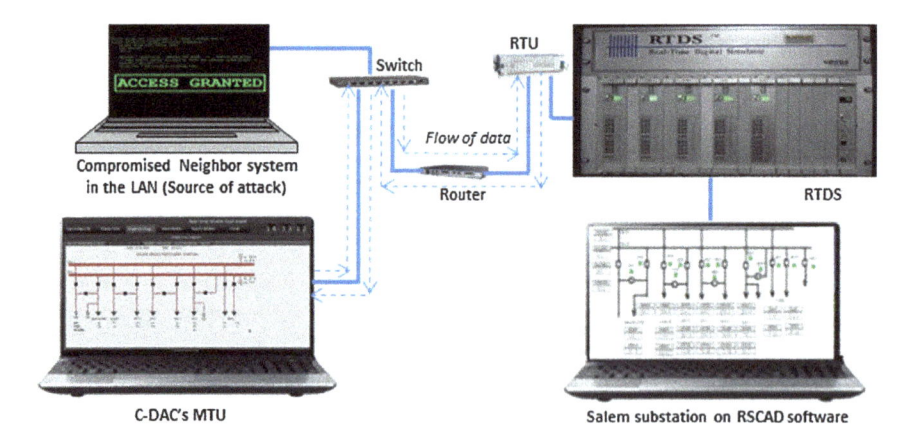

Fig. 2 Experimental setup at CPRI

4.3 Signal Interfacing Between RTDS and RTU

RTU receives the signals from Salem substation which is simulated on RTDS. Analog (real and reactive power flows of incoming and outgoing lines, bus voltages, and frequency) and digital (circuit breaker) parameter signals are connected to RTU from RTDS.

4.4 Environment

C-DAC's SCADA engine consists of an integrated version of a master terminal unit (MTU) and human–machine interface (HMI). MTU collects data from RTU using IEC 60870-5-104 protocol over TCP/IP. HMI is used for data visualization using single-line diagrams. HMI is also used for issuing control commands.

4.5 Attacker's System

The target of attacker's system is to tamper the data between RTU and MTU. This can be achieved using different tools such as Ettercap. Ettercap tool is useful for tamper data in transit and in static. For example, an attacker can use a regular expression to search for a particular pattern of strings and replace text to some other string. Tools such as Ettercap are not suitable in this scenario since communication between RTU and MTU is real time. In a substation, all parameters such as circuit breaker status, load, and frequency are interdependent. In real-time, attacker has to select a tampering value such that MTU should accept the packet. This is one more trivial point why standard tools are not suitable in this case.

In this scenario, we are considering a case of an internal attacker and having all details about the system. Attacker writes a program which can work in obfuscation mode to sniff all the incoming/outgoing data through the switch. After analyzing the data, the attacker can understand the telegram details of data. By exploiting the communication protocol vulnerabilities [10], the attacker can tamper with the payload (telegram's data) for a selected list of telegrams of the packet in such a way that MTU accepts the packet. In this way, the attacker can inject malicious/incorrect values. By exploiting the communication protocol vulnerabilities [10] and understanding the telegram details of digitally controlled devices, the attacker can craft a packet adhering to the standard to control some devices and forward to RTU.

4.6 Plan for Conducting Attacks

Steady-state simulation of the 400-kV Salem substation is carried out in RTDS. There are two types of attacks that are demonstrated in this paper. First one, take the control of the system and start initiating the control commands from attacker machine to bring the system unstable. The second one is tampering the data with appropriate misleading values, so the operator is forced to take wrong decisions.

5 Steady-State Simulation

Steady-state simulation of the 400-kV Salem substation is carried out in RTDS with the power flows matching with the snapshot of the power flows as shown in and Table 1 and as per Salem substation. Single-line diagram of the 400-kV substation represented on RTDS with real-time simulation of bus voltage, active power, and reactive power read by RMS meters is given in Fig. 3.

Table 1 Incoming and outgoing lines of 400-kV substation

	P in MW	Q in MVAR
Incoming lines		
Bangalore to Salem	147	−57
Hosur to Salem	319	−58
NYL-I to Salem	169	73
NYL-II to Salem	175	19
Total power	810	−23
Outgoing lines		
Myvadi-I (UPT-I)	−343	72
Myvadi-II(UPT-II)	−335	79
ICT-I(TNEB)	−18	−21
ICT-II(TNEB)	−111	−35
Total power	−807	95

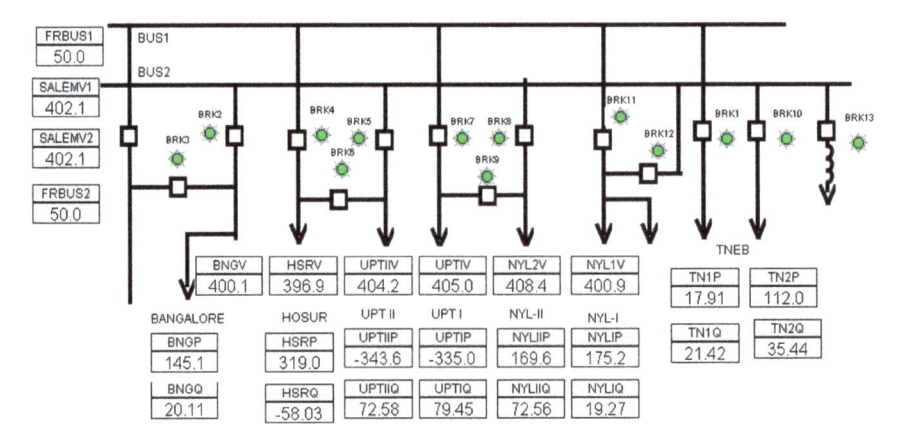

Fig. 3 Steady-state power flows of the system

Analysis of Communication Channel ... 123

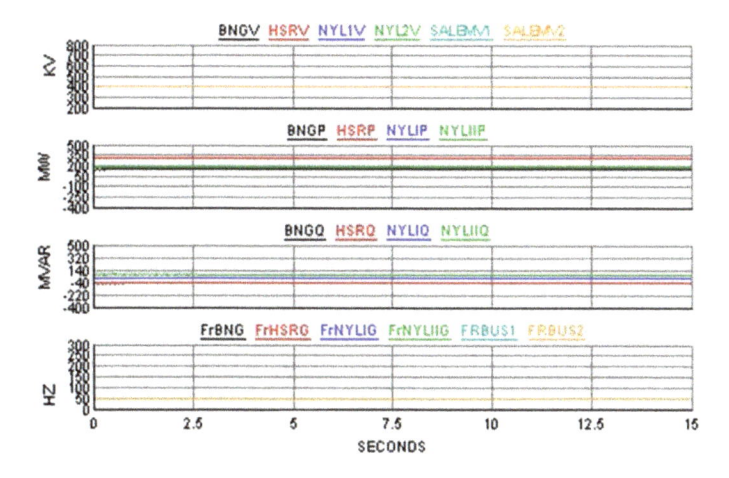

Fig. 4 Steady state—voltages, P and Q flows, frequency on incoming lines

Frequencies, voltages, real and reactive power flows of incoming lines of 400 kV Salem sub-station are shown in Fig. 4. Frequencies, voltages, real and reactive power flows of outgoing lines of 400 kV Salem sub-station are shown in Fig. 5. All the values shown in the Figs. 4 and 5 are steady-state values. Breaker signals from RTU are recorded and shown in Fig. 6. All the breakers signals are indicating high (ON-position) in steady state condition.

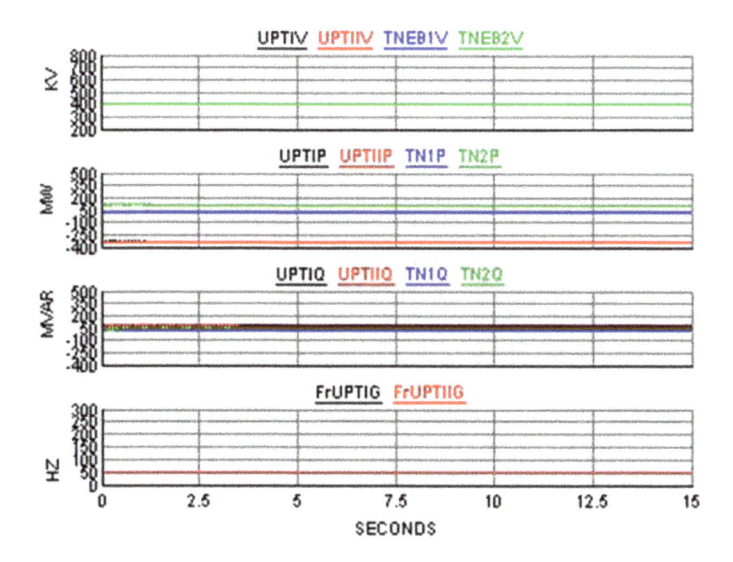

Fig. 5 Steady state—voltages, P and Q flows, frequency on outgoing lines

Fig. 6 Steady-state breaker signals from SCADA system

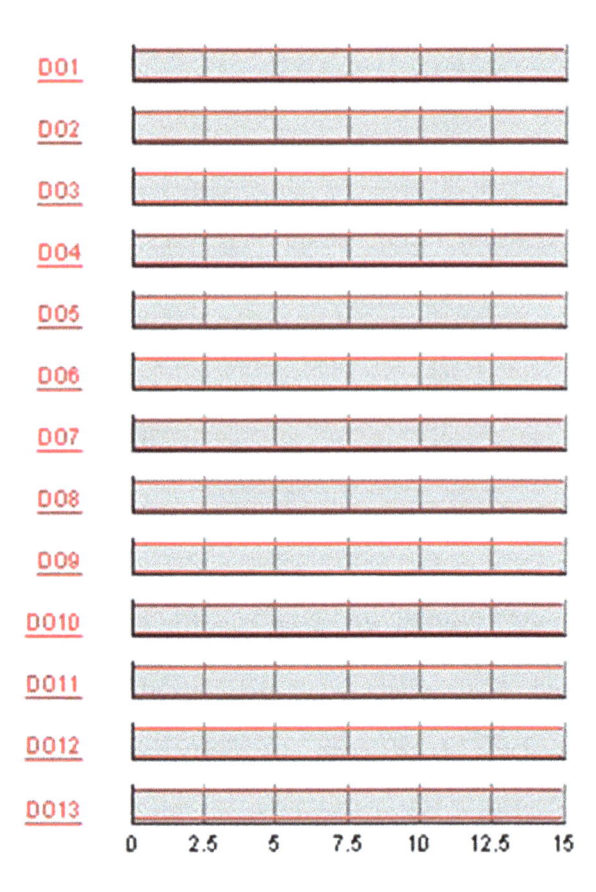

6 Transient Simulation Cases

Transient simulation cases are carried out as attacks by the compromised neighbor system in the same network of MTU. Since tools like such as Ettercap are not suitable in this scenario, C-DAC's in-house hacking tools have been used and the compromised system is successfully able to launch a man-in-the-middle (MITM) attack to route the network traffic between the MTU and RTU through it.

Once it is made sure that the traffic is routed through attacker system, the hacker can either passively look at the data flowing or indulge in data modification, because communication between RTU and MTU is using IEC 60870-5-104 standard protocol. The attacker can also initiate a command to the RTU because a breaker control command C_DC_NA_1 packet structure is as per protocol and it is open. Several attack experiments have been carried out on the simulated 400-kV Salem substation, and the results have been captured using RSCAD software for impact analysis. Among them, two such attack cases have been presented below.

6.1 Attack: Sending Control Command to Hosur Breaker and Tie Breaker by the Attacker

In this attack, the attacker initiates a command (C_DC_NA_1) from his machine as per IEC 60870-5-104 standard. The intention is to bring down the system by controlling few breakers from 'ON' status to 'OFF' status.

Healthy system data (steady-state data):

Hosur source MW: 319 MW, breaker: ON, tie breaker: ON, bus voltage: 396.9 kV, frequency: 50.0 Hz

Line breaker and tie breaker signals of Hosur line are tampered by the attacker and are made ON to OFF by the attacker. With the loss of the incoming power (319 MW) from Hosur feeder, the system is becoming unstable. The system going to unstable condition in these simulation cases is due to the limited simulation of the power system. In an actual system with the loss of one or two lines the stability of the system will not be affected, it is required to perform the same attack many incoming lines or many substations in a coordinated way, which is possible. Single-line diagram of the system with the incoming line from Hosur disconnected is recorded as given in Fig. 7. The bus voltages and frequencies attain abnormal values that can be seen from the figure given below.

After successful conduction of attack1, the system became unstable and all signals are recorded in the same duration of time. Circuit breaker signals from the system are recorded and shown in Fig. 8. Breaker 4 and Breaker 5 signals are becoming 'LOW', and all other breaker signals are 'HIGH'. Abnormal voltages of

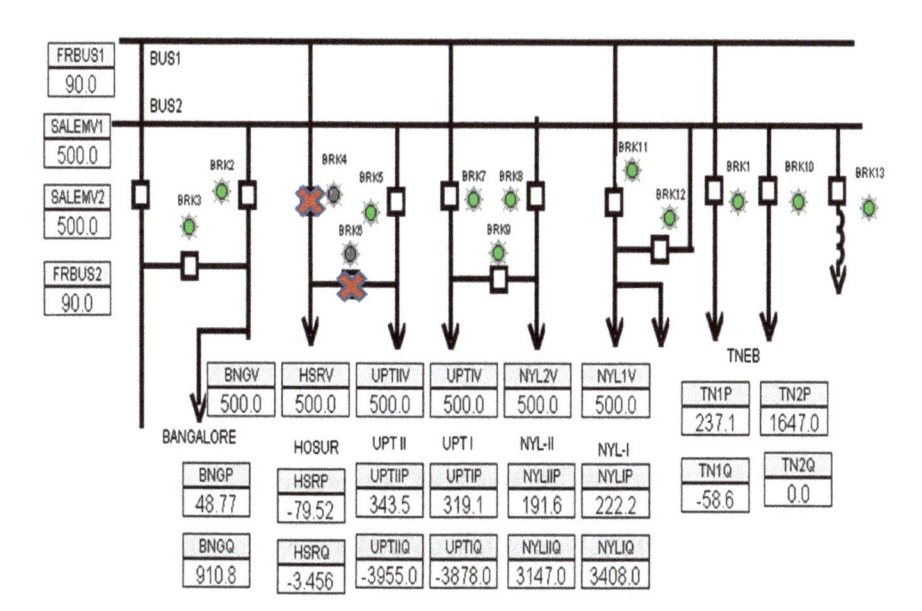

Fig. 7 SLD and power flows for attack1 after incoming power from Hosur is disconnected

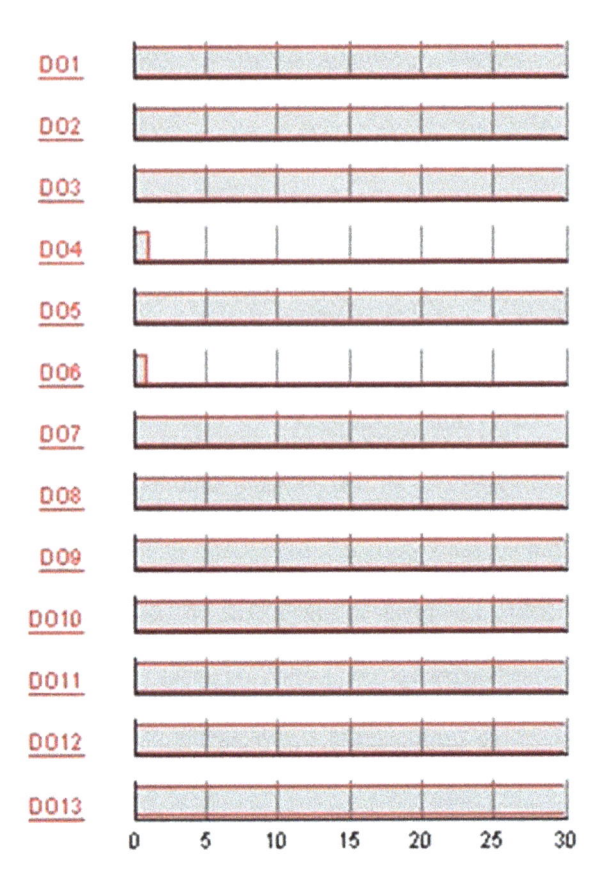

Fig. 8 Attack1—breaker signals from SCADA security system

the incoming lines, real and reactive power flows of incoming lines, and frequencies are given in Fig. 9. Voltages of the outgoing buses, real and reactive power flows on outgoing lines, and frequencies are given in Fig. 10.

6.2 Attack2—Tampering the Load Values (UPT-II & UPT-I), Salem Bus Voltage, and Frequency by the Attacker

Healthy system data (steady-state data):

Myvadi(UPT-II): −343.6 MW, breaker: ON, tie breaker: ON, bus voltage: 404.2 kV, frequency: 50.0 Hz

Myvadi(UPT-I): −335.0 MW, breaker: ON, tie breaker: ON, bus voltage: 405 kV, frequency: 50.0 Hz

Analysis of Communication Channel ... 127

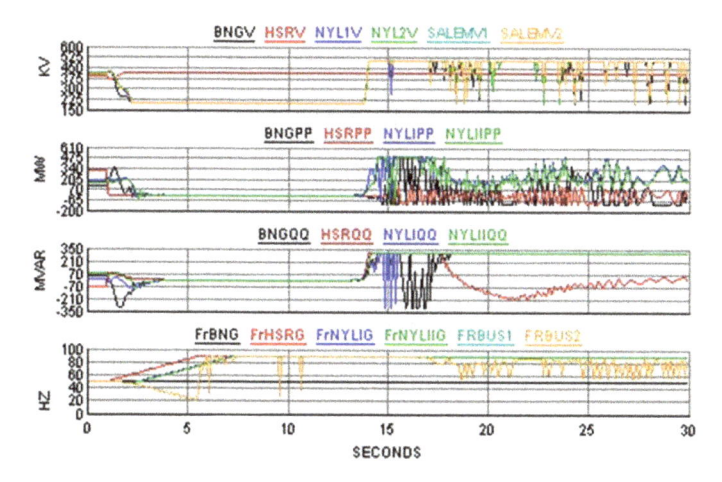

Fig. 9 Attack1—voltages, P and Q flows, frequency on incoming lines

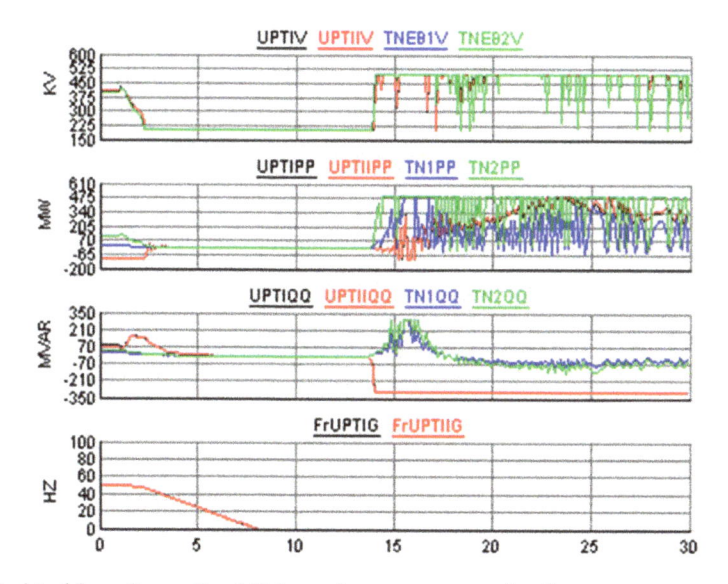

Fig. 10 Attack1—voltages, P and Q flows, frequency on outgoing lines

Salem BUS1 voltage: 402.1 kV, Salem BUS1 frequency: 50 Hz
Salem BUS2 voltage: 402.1 kV, Salem BUS2 frequency: 50 Hz

In this case, the UPT-II and UPT-I power flows, Salem bus voltages, and frequency are tampered by the attacker. Outgoing power from UPT-II and UPT-I is tampered from healthy values of −343.6 and −335.0 MW to unhealthy values of −170 and −160 MW, respectively, by the attacker and sent it to the MTU/operator. Also, Salem BUS1 voltage and frequency are tampered from healthy values of

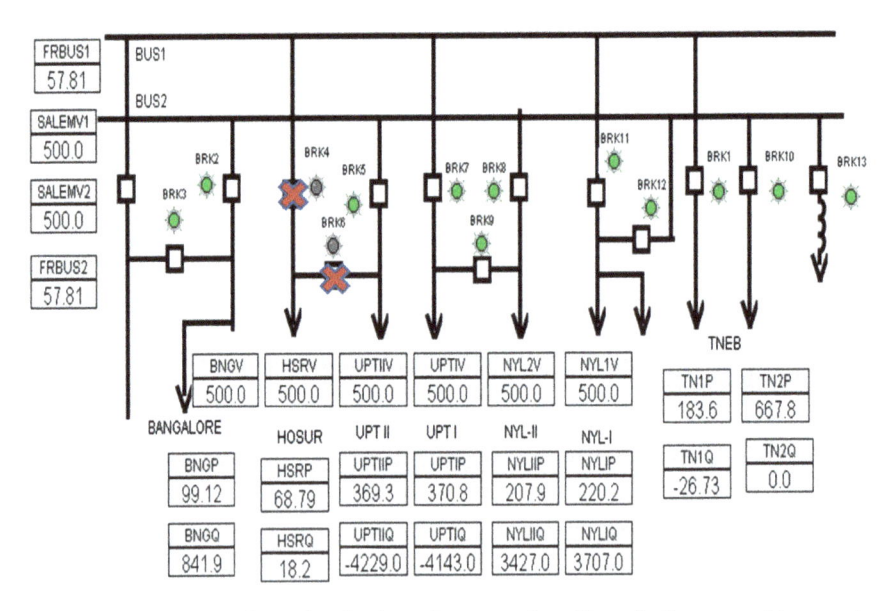

Fig. 11 SLD and power flows for after incoming power from Hosur is disconnected by operator

402.1 kV and 50 Hz to unhealthy values of 440 kV and 51 Hz, respectively, by MITM and sent it to the operator. These tampered values will send to control system for certain duration. Looking at these values, system operator takes the corrective measure and disconnects the most appropriate generation (e.g., Hosur incoming line) by opening the breakers BRK4 and BRK6 to bring back the frequency and voltage. These breaker operations forced the operator after the monitoring of tampered incorrect values. With the loss of the one incoming line, the system is becoming unstable. Single-line diagram of the system with Hosur incoming line being disconnected is recorded as given in Fig. 11. The bus voltages and frequencies attain abnormal values that can be seen from the figure given below.

Breaker signals from the system are recorded and shown in Fig. 12. Breaker 4 and Breaker 6 signals are becoming 'LOW' and shown in Fig. 12 showing the attack, and all other breakers signals are 'HIGH'. Parameters of the incoming and outgoing lines are going abnormal that can be seen from Figs. 13 to 14. The system became unstable in case operator forced to perform the unwanted control operation by monitoring tampered incorrect system state.

Analysis of Communication Channel ...

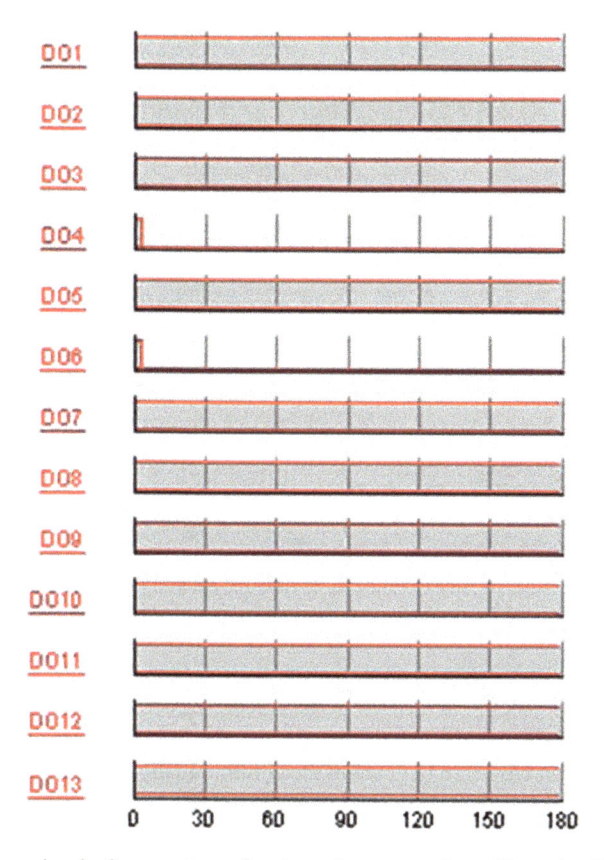

Fig. 12 Breaker signals from system after incoming power from Hosur is disconnected by operator

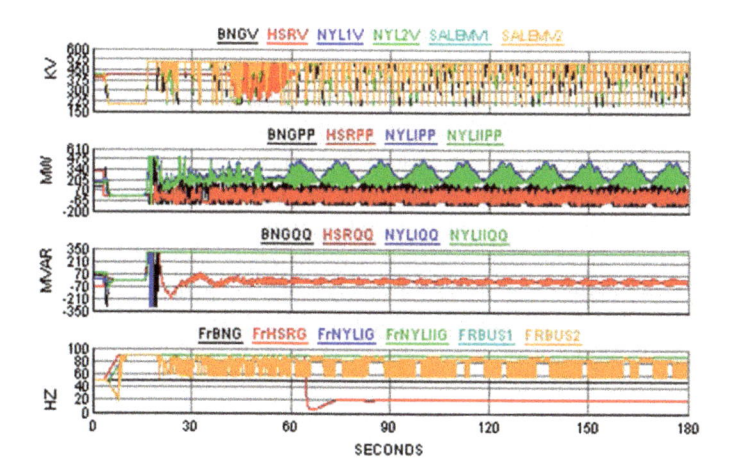

Fig. 13 Voltages, P and Q flows, frequency on incoming lines after incoming power from Hosur is disconnected by operator

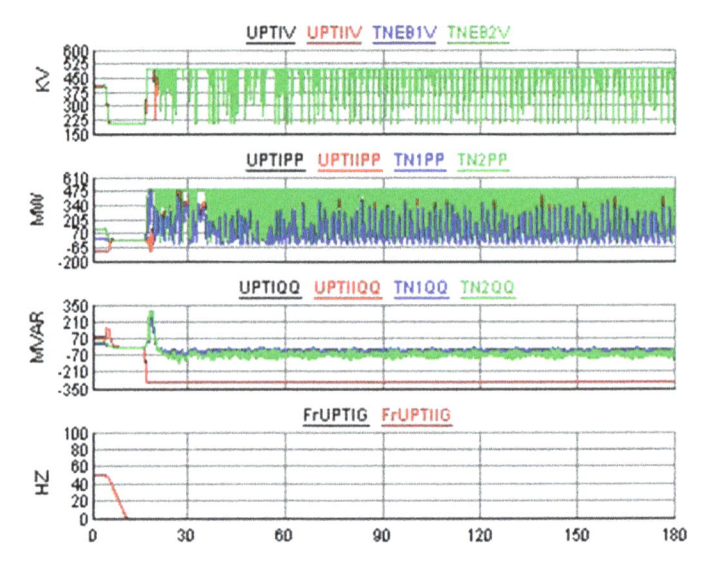

Fig. 14 Attack2—voltages, P and Q flows, frequency on outgoing lines after incoming power from Hosur is disconnected by operator

7 Countermeasures for Attacks

Transient simulation cases are targeted at the communication channel between RTU and MTU. Targeted protocol in this scenario is IEC 60870-5-104 which is a plaintext protocol used for communication. To harden the data flow between RTU and MTU, this protocol needs to be hardened. In SCADA, availability is the first preferred criterion over integrity and confidentiality. Interoperability also plays a key role in SCADA systems, since devices need to communicate with third-party systems. Keeping availability and interoperability in consideration, standard information technology practices like encryption or proprietary authentication mechanisms cannot be implemented.

To harden IEC 60870-5-104 protocol, the recommended standard is IEC 62351 [11]. IEC 62351 addresses various vulnerabilities in the protocol and defines a mechanism for data exchange. It addresses application-layer security and also mitigates attacks such as man in the middle, replay, data modification, and non-repudiation. Make sure that implementation of IEC 62351 over IEC 60870-5-104 does not affect the availability of the application. IEC 62351 defines the mechanism of authentication, key exchange, etc. By adhering to this, interoperability will not be affected since any other device adhering to this open standard can communicate each other.

8 Conclusion

SCADA networks are vulnerable to cyber attacks. With the adaptation of open standard protocols and when these protocols are not designed with security in mind, the threat of exploiting these networks by attacker increases by multifold. IEC 60870-5-104/101 protocols are widely used in power sector in India, and these protocols are not designed with security in mind. Any successful attack would lead to catastrophic effect. SCADA systems do not behave in the same way as IT systems. To deploy countermeasures, a preliminary requirement is to study the impact analysis of attacks. For the same, it is not possible to conduct attacks in real time. In this paper, we have set up an experimental platform using C-DAC's SCADA system and RTDS. Multiple experiments have been conducted to make the 400-kV Salem substation unstable by exploiting the vulnerabilities in IEC 60870-5-104 protocol, and impact has been analyzed. So, one needs to be proactive to study the vulnerabilities and analyze the risk involved and implements appropriate security steps.

References

1. https://en.wikipedia.org/wiki/Stuxnet
2. https://en.wikipedia.org/wiki/December_2015_Ukraine_power_grid_cyberattack
3. https://en.wikipedia.org/wiki/Duqu
4. IEC standard for IEC 60870-5-101 protocol
5. IEC standard for IEC 60870-5-104 protocol
6. http://www.cpri.in/about-us/departmentsunits/power-system-division-psd/real-time-digital-simulator.html. Last accessed on 29 July 2016
7. Amaraneni A, Lagineni M, Kalluri R, Senthil Kumar RK, Ganga Prasad GL (2015, March) Transient analysis of cyber-attacks on power SCADA using RTDS. J CPRI 11(1):77–80
8. Samanth DP, Kalluri R, Senthil Kumar RK, Bindhumadhava BS (2012) SCADA systems security: threat analysis using defense graphs. In: International conference on physical, cyber & system security for power sector from 27th to 28 Feb 2012
9. Sommestad T, Ekstedt M, Nordström L (2009) Modeling security of power communication systems using defense graphs and influence diagrams. IEEE Trans Power Deliv 24(4)
10. Samanth D, Kalluri R, Senthil Kumar RK, Bindhu Madhava BS (2013) SCADA communication protocols: vulnerabilities, attacks and possible mitigations, CSI transaction on ICT 1.2, pp 135–141
11. IEC security standard—IEC 62351

Plug and Operate Power and Distribution Transformer Technology

Deepal Shah

Abstract The paper gives details of innovative smart Plug and Operate Transformer technology. It makes transformer touch-proof, shockproof, and maintenance-free. The transformer remains sealed in all condition while fixing or replacement of bushing, and hence, possibility of moisture is minimized. Technology helps to restore power in less than 3–4 working days and substantially minimizes failure of transformer.

Keywords Plug-in concept in HVAC transformer bushing · Plug-in concept in HVAC cable terminations · Plug-in concept in dry-type surge arrester

1 Introduction

Our power distribution network is changing. Urbanization leads to new structures integration of renewables require an updated concept for network infrastructure. Compactness and size reduction are no longer a feature, but crucial. One approach is to use a plug-in, dry-type high-voltage connection system which supports the easy, fast, and safe expansion and reconstruction of the power grids. Downtime has to be minimized, whereas safety aspects become more complex.

In addition to this, electric power supply systems are—beneath the normal operating voltage—often exposed to overvoltage events. The voltage level during these events may go near or even over the limits of the main insulation. Especially transient overvoltage is critical for the insulation of high-voltage equipment. To prevent damage, surge arrestors are used to limit overvoltage to a voltage level. In this paper, the differences and equalities of pluggable and conventional surge arrestors are discussed as well as possibilities for pluggable terminations.

D. Shah (✉)
PFISTERER Holding, Winterbach, Germany
e-mail: Deepal.shah@pfisterer.com

© Springer Nature Singapore Pte Ltd. 2018
R. K. Pillai et al. (eds.), *ISGW 2017: Compendium of Technical Papers*, Lecture Notes in Electrical Engineering 487, https://doi.org/10.1007/978-981-10-8249-8_12

2 Ease of Use

Such a connecting system consists of two main parts, the socket and the connector, shown in Fig. 1.

The socket is fitted into a switchgear or transformer. It encloses the equipment and ensures its tightness. Installation of the socket offers the biggest advantage if installed during manufacturing.

Different options are available for the socket, either connecting to an overhead line using a plug-in bushing, connecting an XLPE cable using a connector or connection of a pluggable surge arrester. Furthermore, the system can be enclosed by a dummy plug for future use.

The separable connector which is plugged into the socket consists of the contact element, an insulating part including the electrical stress relief, and a metallic housing.

The dry pluggable CONNEX system can handle cable diameters of 3000 mm^2 CU or AL conductor, carrying up to 4000 A nominal current. Maximum diameter over insulation is 144 mm. Principal setup is shown in Fig. 1. An additional benefit offers the housing, which is completely touch-proof and waterproof. It is saltwater resistant and therefore can be used in coastal areas as well as in real offshore surroundings (Figs. 2 and 3).

SF6 usage is getting more and more restricted [3, 4], and testing requires SF6 gas handling. Whereas air insulation requires creepage and space, furthermore positioning is not optimal as substation grid layout usually does not allow the

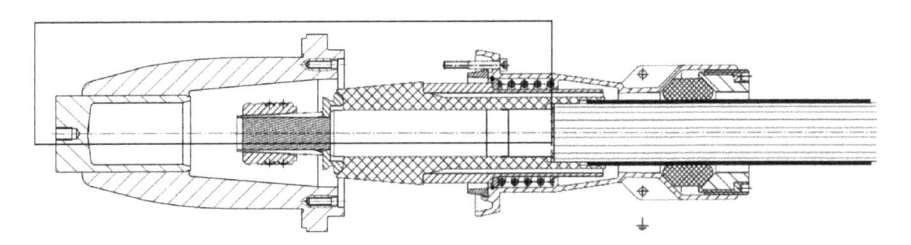

Fig. 1 Socket including HV stress relief (not shown)

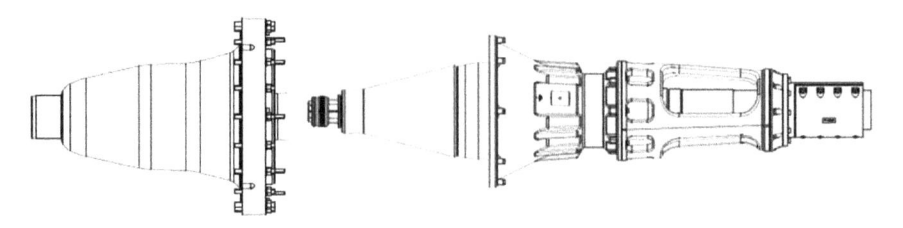

Fig. 2 Example of a pluggable system U_M = 550 kV

Fig. 3 Dry-type plug-in bushing

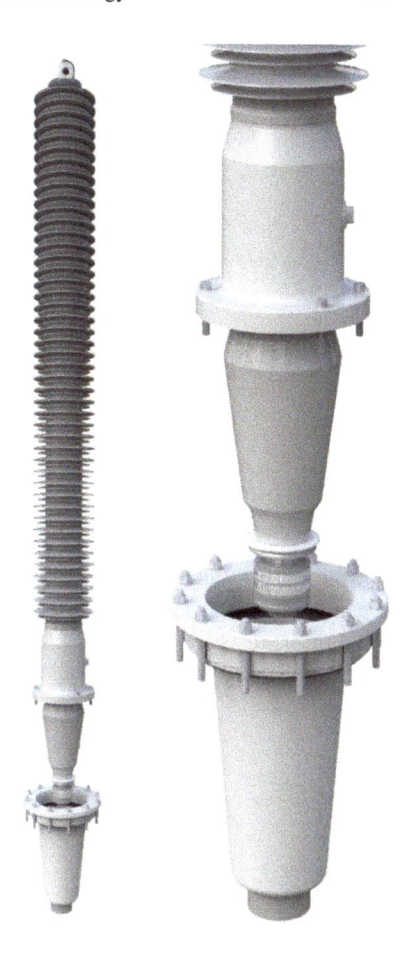

air-insulated arrester to be positioned as close to the transformer as possible when cables are being used.

A pluggable solid-insulated arrester has been developed and has been tested up to $U_c = 144$ kV ($U_r = 180$ kV) in reference to the latest standards.

The arresters main insulation is solid; there is no insulation liquid or insulation gas such as SF6 included. Figure 4 shows a cut view of such an arrester. The main part, regarding the function of a surge arrestor, is metal oxide resistor tablets. These MO tablets are used as a nonlinear component with a very low leakage current during operation. The tablets are connected to the male part of the plug-in system and are insulated by a silicone body. This insulating body includes field-controlling elements. The head armature includes a bursting disk for pressure relief and a turnable head for the redirection of the gas, in the event of a failure according to IEC 60099-4. The housing is made of glass fiber-reinforced resin and allows enormous mechanical strength as well as protection of the silicon body against environmental conditions. The silicone body itself is protected and

Fig. 4 Dry-type plug-in surge arrester

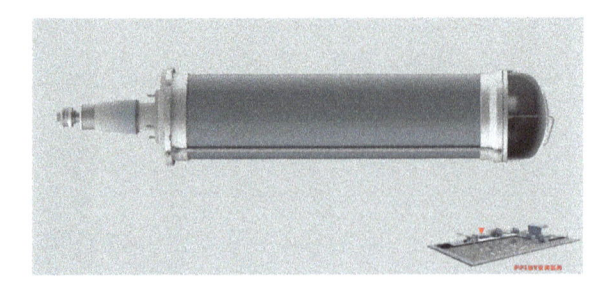

touch-proof. A special arrangement of the earthing path allows connecting monitoring devices or discharge counters if desired.

Pluggable solid-insulated surge arresters therefore combine all the benefits of the surge protectioning of a GIS arrester, and the advantages of a plug-in system are:

- Minimum installation distance and thus maximum electrical protection of the equipment, especially for compact and touch-proof transformers.
- No pressure vessel, no gas, and no monitoring system necessary because of the usage of a dry insulation of elastomer.

Part II
E-Mobility

Public Opinion on Viability of xEVs in India

Mohd. Saqib, Md. Muzakkir Hussain, Mohammad Saad Alam, M. M. Sufyan Beg and Amol Sawant

Abstract This work demonstrates a smart charging system of electric vehicle using information technology and cloud computing. xEVs (electric plugin hybrid, battery electric vehicles) charging management system will be very helpful for the varying charging infrastructure demands, namely perspectives from automakers, electricity providers, vehicle owners, and charging service providers. Through dedicated interface, the developed system will provide real-time information to xEV users regarding nearest charging station with minimum queuing delay and with minimum charging cost through a secured online accessing mechanism for accessing state of charge (SOC) of the xEV's battery being charged. The system not only provide an execution framework for the xEV users but also provide an optimal energy trading solution to all entities involved in a smart charging infrastructure such as charging station, aggregators, smart grid. The work also explains the cloud-enabled bidding strategies that look for day-ahead and term-ahead markets. The aggregators will use the smart decisions undertaken by cloud analytics to execute their bidding strategies in way to maximize the profit. Further, the work also assesses the possible cybersecurity aspects of such architectures along with providing possible solutions.

Mohd. Saqib · Md. M. Hussain · M. S. Alam (✉)
Centre of Advanced Research in Electrified Transportation,
Aligarh Muslim University, Aligarh, India
e-mail: saad.alam@zhcet.ac.in

Mohd. Saqib
e-mail: saqib9828@gmail.com

Md. M. Hussain
e-mail: md.muzakkirhussain@zhcet.ac.in

M. M. S. Beg
Department of Computer Engineering, Muslim University, Aligarh, India
e-mail: mmsbeg@eecs.berkeley.edu

A. Sawant
India Smart Grid Forum (ISGF), New Delhi, India
e-mail: amol.sawant@indiasmartgrid.org

© Springer Nature Singapore Pte Ltd. 2018
R. K. Pillai et al. (eds.), *ISGW 2017: Compendium of Technical Papers*, Lecture Notes in Electrical Engineering 487, https://doi.org/10.1007/978-981-10-8249-8_13

Keywords Indian energy exchange · Electric vehicles (EVs) · Cloud computing
Energy price · Energy storage · Energy markets · Information technology
Cyber security · Smart grid

1 Introduction

The rapid urbanization and industrialization are posing pollution issues as the prime concern in India. Automotive sector constitutes the main sources of such pollution. Air pollution is often understood as an urban issue [1], and it is appealing that cities concentrate economic activity and demand for energy services and so tend to experience heavy concentrations of harmful air pollutants. Electric vehicles [2] potentially emerged as supreme player in realizing an eco-friendly, pollution-free means of next-generation transport [3]. Due to venturing features associated to electric vehicles, they are acquiring a consensus call from automotive industries, R&Ds, policymakers contrast to their ICE counterparts.

The rollout of xEVs is an ongoing objective for all developing nations including India and provides numerous nascent thrusts for utility providers and industrial sector [4–6].

Heavy penetration of xEVs will create enormous demand sink for electrical energy needed to charge xEVs. An uncoordinated charging infrastructure will create sudden demand upsurges, put constraints on the underlying power grid and at the worst case may lead to failures and outages [7]. The demand of electrical power will be different at different charging stations according to number of xEVs deployed or routed per station. However, all the charging stations require strict timing constraints in their charging patterns so that the xEVs could be charged service reliably [8]. As the charging stations are coupled to the main supply system, the power drawn by them in different time instants should not induce encumbrance on the back end smart grid (SG) system [9]. Thus, it is the call of the hour to develop a centralized EV/PHEV charging management system which controls demand and supply problem within the dedicated time frames.

Meanwhile, the xEVs, emerged as a new kind of power load, could perform active role in diverse energy market services such as vehicle to grid (V2G) [10], frequency regulation [11], ancillary services [12] and would potentially exert an incredible impact on the daily residential load curve if they are properly managed [13]. But such abilities latent in the xEV fleet could only be brought to commercial use if their charging and discharging methodologies are intelligently tuned [14].

Thus, there is a need for benchmarking efficient management protocols and standards to ensure that the fleets are regulated smartly [4].

A master coordination framework having reliable service-oriented software architecture (SOA) should thus be developed that provides a platform for the xEV users which enable them to participate in regular and auxiliary energy market operations [15]. Such platform is analogous to power exchanges (PXs). It feeds the daily demand and supply trends on its database and correspondingly settles a proper

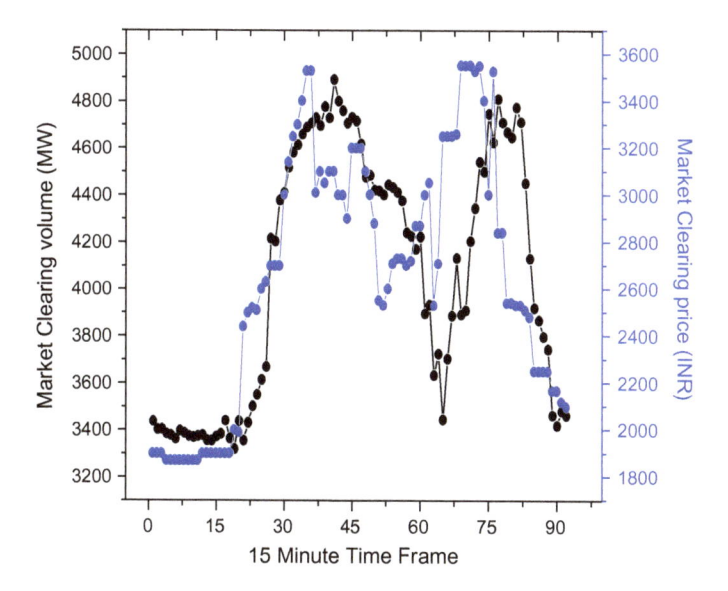

Fig. 1 Load and price curve for a typical power market

market clearing price for the involved stakeholders such that the interest of each player is respected in a win-win paradigm. Figure 1 shows the snapshot of one-day market clearing volume corresponding to stipulated clearing price. Cloud computing has recently emerged as technology enabler for the smart cities, smart health, smart transportation, and smart environment as well as for smart gird [16].

Cloud computing utilities provide massive virtualized storage spaces which can be deployed in pay-per-use models to realize warehouse or database needs of such software. Contemporary developments in the intelligent transportation system (ITS) aided with utilities for information and communication technology (ICT) have enabled bidirectional data collection and transport mechanics where the real-time vehicle data can be utilized for modeling fleet charging management prototypes [17].

Motivated by the facets of cloud computing paradigms and the need for an efficient frame work for coordinated charging of xEVs, this work proposes centralized cloud-based remote xEV charging management system. Figure 2 depicts the integrated xEV data-aware fleet management architecture constituted by data and energy trade mechanisms among the acting entities. Inheriting the ideology of duplex power and data exchange services, the work demonstrates prototyped software that fed xEV customer parameters such as current geographical location and state of charge of xEV battery and correspondingly guides him to appropriate charging station.

The application acts analogous to an energy exchange (EX) that accumulates the bids from multiple vendors and aggregators and at the same instant extracts the day-ahead or term-ahead demand market attributes, for realizing a customer-centric

Fig. 2 Data and energy flow for xEV charging management

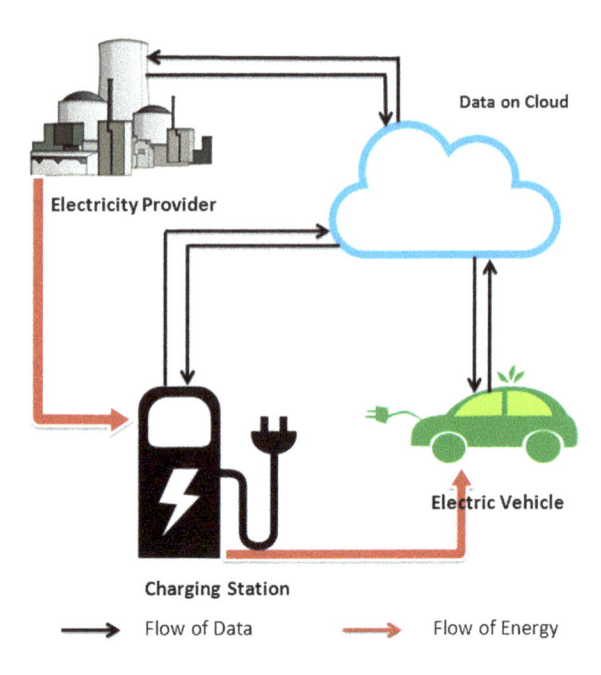

recommendation environment. The work also presents the software specifications required to prototype similar applications.

The detailed contributions are summarized as follows:

1. Proposed a cloud-based remote xEV charging management framework recommending optimal charging solution to the involved xEV customers.
2. Proposed a prototypical software description for developing such applications.
3. The ER diagram, use case diagram, and dataflow diagram for a typical xEV charge management software are developed that will support a consistent standard for the programmers and application developers.
4. Highlighted the cyberthreats that may arise while installing such applications and also addressed the need of security and integrity enforcements while running such applications.

The manuscript is organized as follows. Section 2 explains operational business model for the involved entities. Section 3 describes the software description for developing prototypical smart charge management system. In the fourth section, the working of the application is depicted along with the cyberissues that may arise in course of application execution. Section 5 concludes the paper.

2 Business Model

This proposed system, xEV charging management system, demonstrates the significance of cloud-based paradigm focused to deploy a smart and coordinated charging management framework for xEV fleet. The front end of the software is developed using ASP.NET to have better GUI realization for system end, while for backend operations, C#.NET is used.

An instance of data center is developed using SQL Server 2008 purposed to manage the computation and analytics tasks such as login (authentication), searching, updating information regarding energy attributes, and bidding analysis. The proposed system demonstrates the bid and auction mechanism under took by power system stackholders. It exposes a virtual marketplace for the electric vehicle users who track the real-time energy market attributes such as present power rate predicted power flow and cost. The cloud analytics undergoing at the data center employees computationally perceptive be data analytics to recommend smart charging options to the corresponding vehicles customer. The application supports dedicated interfaces to provide real-time information to the xEV user regarding nearest charging station that can provide energy at economy rate and that so by ensuring a minimum queuing delay.

The application not only acts as a recommendation engine for the xEV users but also provides an optimal charging solution to all entities involved in a smart charging infrastructure such as charging station, aggregators, smart grid.

Charging stations manage their user through this application-performing task such as updating charging level, places to stay, update list of selected consumer after bidding. This application reduces load on smart grid by solving the problem of demand and supply problem.

3 Software Description oF Proposed Model

This section presents the detailed software description of the prototype system using different models for representing interaction among entities, the schema description, the control and work flow among the involved actors, etc.

3.1 ER Diagram

An entity relationship diagram (ERD) represents the relationships among entity sets stored in the database. An entity in this context is a component of data. In other way to say, entity relationship diagrams illustrate the logical configuration of databases. There are three main entities involved in the model:

(1) **Admin**: The system operator (admin) entity has two major properties, Admin_ID and password, through which to login, and then, admin can perform other task (Fig. 3). For the proposed application, the administrative legacy will be licensed to different stakeholders based on the registration and legal procedures.
(2) **Power Plant Operator**: The power plant operator (PPO) or independent system operator (ISO) registered by admin with a unique PP_ID and a

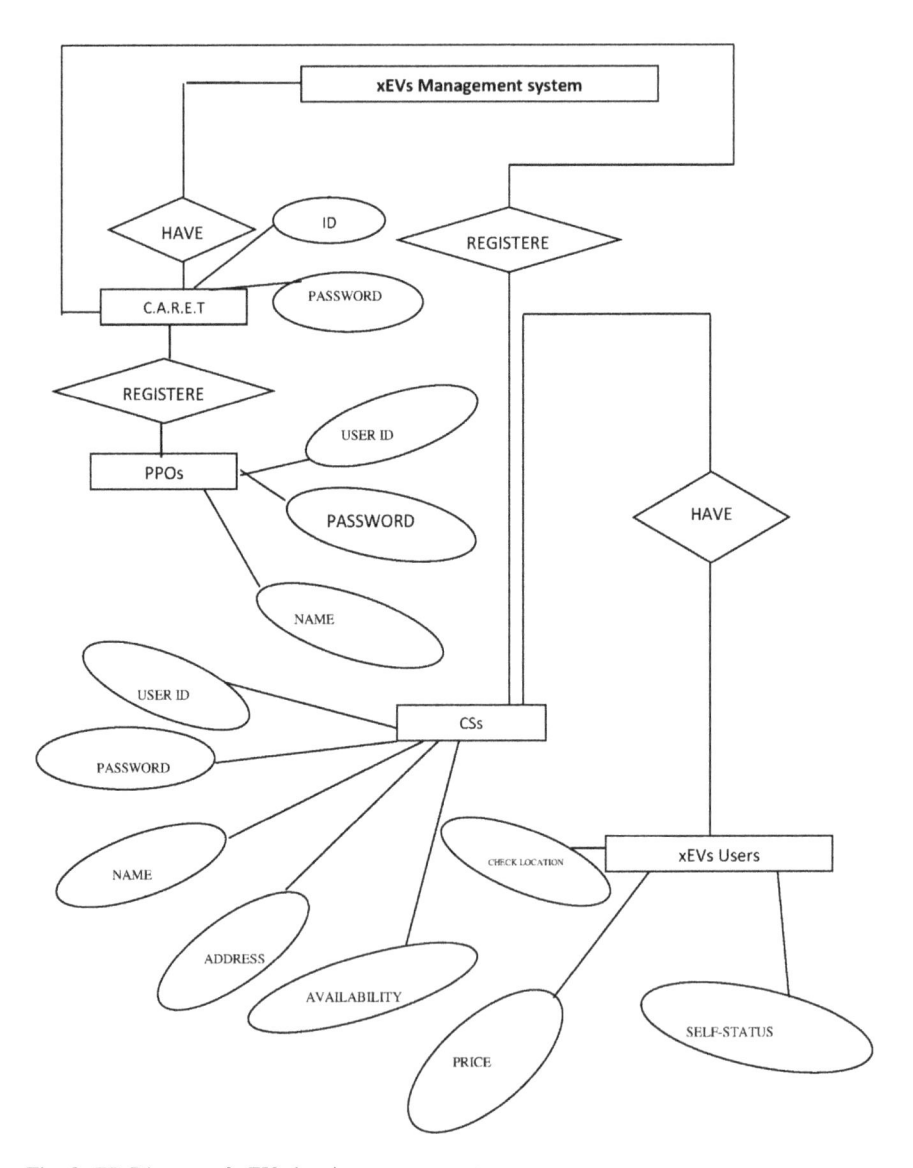

Fig. 3 ER Diagram of xEV charging management

self-assigned password which can later be changed by power plant operator. Further attributes of power plant operator are name, location (updatable), contact details (address, email, phone number), and energy transformation related (current price, working/transformation time, capacity, etc.).

(3) **Charging Station Operator**: The charging station operator (CSO) also registered by admin with a unique CS_ID and a self-given password which can later change by charging station operator. Further properties of charging station operator are name, location, contact details (address, email, phone number), energy transformation related (current price, working/transformation time, capacity, etc.), and a record of nearby places where xEV user can stay.

(4) **xEV User**: The xEV user can register themselves using an online form filling process. After admin approval, its session is assigned through a unique USER_ID.

xEV users have further properties like name, contact details (address, email, and phone number), xEV vehicle identification number (VIN), and type of xEV battery.

3.2 Use Case Diagram

Use case diagrams also referred to as behavior diagrams are used to describe a set of actions (use cases) that some system or a batch of systems (subject) will or can perform in collaboration with one or more external users of the system (actors). In other words, it represents to which functionality(s) the actors/entities in the system involved for. Each use case should provide some observable and valuable result to the actors or other stakeholders of the system.

For the proposed framework, in correspondence with xEV user, the application will utilize the key attributes such as GPS location of xEVs, the state of charge (SOC) and state of health (SOH) of xEV battery, and other miscellaneous attributes to calculate the electric vehicle range anxiety (EVRA) in percentage as shown in Fig. 4.

(a) The application will be able to track status or level of charging (SOC) of their own xEV through the application.
(b) Booking and bidding for the energy at minimum tariff.
(c) It can identify location of optimal charging station, as the system will recommend a set of options in decreasing order of priority.

The charging station operator (CSO) is the second-level user who gets control over the consumers and self-database and also has seamless communication with smart grid operators followed by authentication from the admin. This entity can perform the tasks.

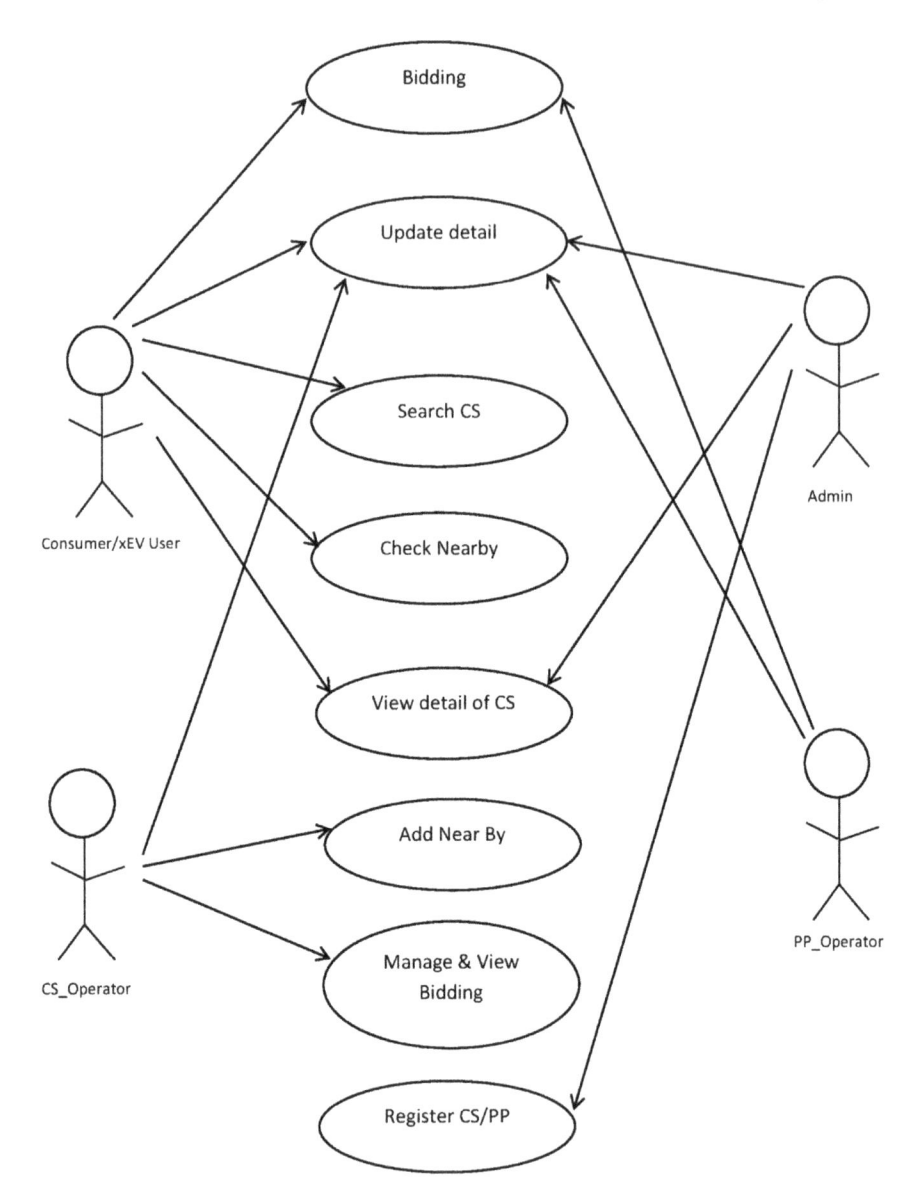

Fig. 4 Use case diagram for proposed model

(a) Control over self-database such as update price for consumer and update demand for main suppliers.
(b) Management for the xEVs such as give notification when EV charged and give availability.
(c) Fix the lowest price for bidding and final the price to buy from main supplier and sell to the consumer.

Power Plants benefits operator is the main supplier of energy who has various tasks after login such as:

(a) View data of CSs such as consumers, average expenditure, location.
(b) Management of self-records such as update energy price, availability of energy.
(c) View and sell energy after bidding.

3.3 Data Flow Diagram [18]

The data flow diagram (DFD) illustrates the graphical depiction of the flow of data among the entities of software. In structural design of software modeling, a DFD can visualize the data processing steps. At the xEV user end, the application will take state of charge (SOC) and state of health (SOH) of xEV battery, the current geographical coordinates of the xEV, and the degree of EVRA as input and provides the optimal location of the charging station as recommended output. Further, when the user is a charging station vendor, the application will feed the bidding volume, bidding price, and the geo-distribution of xEVs in the vicinity as input and engages customers so as to maximize its profit Fig. 5.

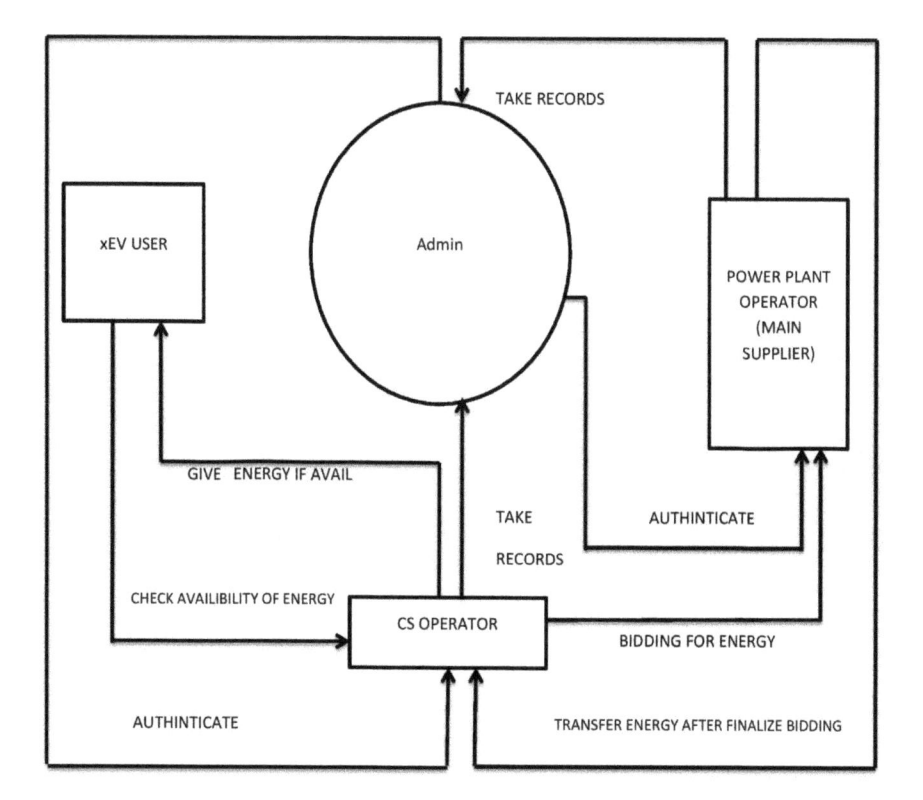

Fig. 5 DFD for the proposed model

4 Proposed System Output Parameters

The application can be deployed in various operating environments such as smartphones, charging station outlets, servers. Based on the portability needs of the execution platform or operating system, the application can be augmented with compatible updates or programming reorganization [19]. However, the software will act as a standard to open doorway to software developers to program applications dedicated to smart grid utilities and services. Figure 6 shows a snapshot of the pilot application while running at the xEV user end.

Moreover, the framework may also be prone to the cyberthreats, thus causing in cyberphysical threats for the overall smart grid ecosystem. Since the underlying smart grid system that is carrying the xEV fleet is a complex system, comprising of massive hardware components requires pervasive control and monitoring infrastructures. The cloud-aware technologies play fundamental role underpinning varying smart grid functionalities, to automate remote management tasks such as xEV fleet management, perform event-driven processes, etc. Such intensive use of cyberphysical infrastructure poses serious threats with respect to privacy, authentication, intrusion, and novelty of such framework [20]. Any breach in such a delicate system from cyberfabric may have serious consequences such as power interrupt, chaos in the fleet due to mismanagement, data breach, accidents.

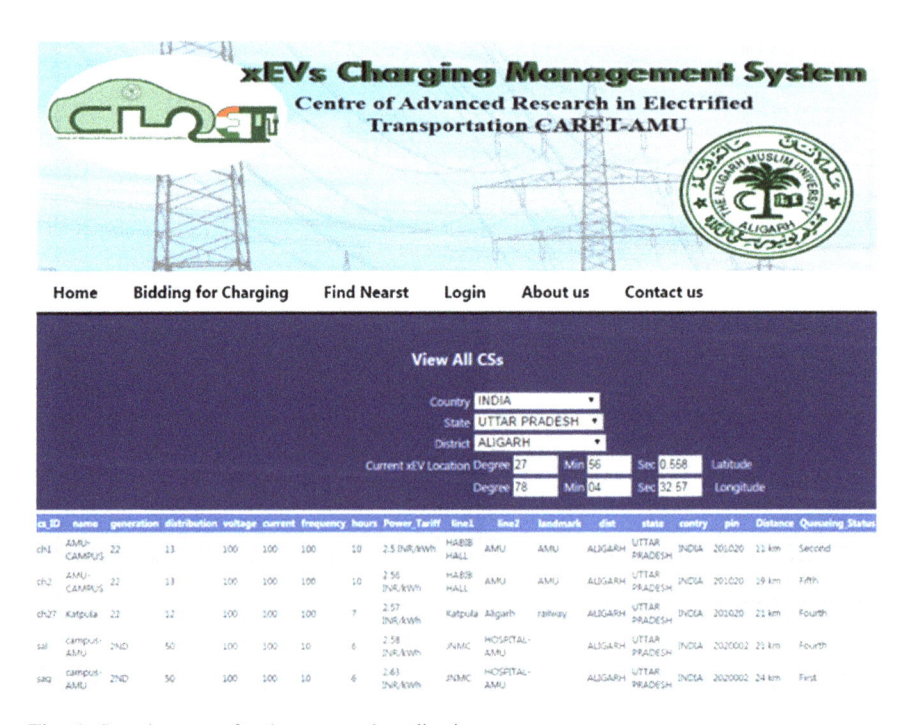

Fig. 6 Sample output for the proposed application

Therefore, a holistic and secure cyberframework needs to be deployed to address the possible vulnerabilities presented in varying modes of execution of the application.

5 Conclusion

The proposed work presents a cloud-aware smart charge management framework for intelligently charging the xEV fleet. The application provides stakeholder-specific interfaces and modules to enable an uninterrupted power flow across the whole smart grid infrastructure and simultaneously circumventing the demand peaks created by the xEV penetration. When installed at the xEV customer end, the cloud-monitored software will be able to track the geo-distribution of optimal charging station that makes the multi-objectives meet concurrently. Among the list of recommending parameters, the application will specifically focus on the three requirements in decreasing order or priority, namely minimum charging tariff, nearest charging spot, and the station having minimum xEV congestion. Moreover, the system will also support dedicated services for charging vendors, aggregators, and other market players. The application will track the status of day-ahead and auxiliary market to settle effective bidding and market clearing environment that maintains a win–win strategy for every stakeholders. The work also addresses the need of cyberphysical enforcements to ensure hassle-free operation to complete SG infrastructure.

References

1. International Energy Agency (2016) Energy and air pollution. World Energy Outlook—Spec. Rep., p 266
2. Sher HA (2012) Addoweesh KE Power storage options for hybrid electric vehicles-A survey. J Renew Sustain Energ 4(5):1–22
3. Pand H (2015) Catch 22 : electric vehicles and the required infrastructure
4. Fang X, Misra S, Xue G, Yang D (2012) Smart grid-the new and improved power grid: a survey. IEEE Commun Surv Tutorials 14(4):944–980
5. Hannan MA, Azidin FA, Mohamed A (2014) Hybrid electric vehicles and their challenges: a review. Renew Sustain Energy Rev 29:135–150
6. Bae S, Kwasinski A (2012) Spatial and temporal model of electric vehicle charging demand, 3(1):394–403
7. Xu J, Wong VWS (2011) An approximate dynamic programming approach for coordinated charging control at vehicle-to-grid aggregator. IEEE Int Conf Smart Grid Commun (SmartGridComm) 2011:279–284
8. Saad Alam M (2013) Key barriers to the profitable commercialization of plug-in hybrid and electric vehicles. Adv Automob Eng 2(2)
9. Bayram IS, Michailidis G, Papapanagiotou I, Devetsikiotis M (2013) Decentralized control of electric vehicles in a network of fast charging stations. In: Globecom 2013-Symposium Sel Areas Communication (GC13 SAC), pp 2785–2790

10. Wu C, Member S, Mohsenian-Rad H (2012) Vehicle-to-aggregator interaction game. IEEE Trans Smart Grid 3(1):434–442
11. Wang Q, Liu X, Du J, Kong F (2016) Smart charging for electric vehicles: a survey from the algorithmic perspective. IEEE Commun Surv Tutorials 18(2):1500–1517
12. Chan CC, Jian L, Tu D (2014) Smart charging of electric vehicles—integration of energy and information. IET Electr Syst Transp, pp 1–8
13. Kang Q, Member S, Wang J, Zhou M, Ammari AC (2016) Centralized charging strategy and scheduling algorithm for electric vehicles under a battery swapping scenario. IEEE Trans Intell Transp Syst 17(3):659–669
14. Eurelectric (2015) Smart charging: steering the charge, driving the change, p 57
15. Yang Y, Wu L, Yang S (2012) The structure of intelligent grid based on cloud computing and risk analysis, In: Elligent human-machine systems and cybernetics (IHMSC), 2012 4th international conference, vol 2, pp 123–126
16. Aburukba R (2015) Role of internet of things in the smart grid technology. J Comput Commun 3:229–233
17. Alam MS (2015) Vehicle to Cloud (V2C) remote management of electric, hybrid and plug-in hybrid electric vehicle charging.pdf. p. Patent Application # 34590/DEL/2015
18. Elbakush A, Functional Modeling with Data Flow Diagrams, Utrecht University. Available at www.students.science.uu.nl/~5771668/me/AssignmentD_5771668_Elbakush.pdf
19. Rawat DB, Rodrigues JPC, Stojmenovic I (2015) Cyber-physical systems: from theory to practice. CRC Press. ISBN: 9781482263329 - CAT# K24361
20. Sharif H (2014) A survey on cyber security for smart grid a survey on cyber security for smart grid communications. IEEE Commun Surv Tutorials 14(4)

Technical and Economic Feasibility Analysis for deployment of xEV Wireless Charging Infrastructure in India

Vatsala, Raqib Hasan Khan, Yash Varshney, Aqueel Ahmad, Mohammad Saad Alam and Rakan C. Chaban

Abstract In the present scenario, the inexistence of commercially viable charging infrastructure is one of the major obstacles in the deployment of plug-in hybrid electric vehicles (xEVs) or wirelessly charged EVs in India. Commercially available EVs can be charged through conductive (plug-in) or inductive (wireless) charging methods. The current automotive industry throughout the globe considers wireless charging methods for addressing the range anxiety of EV customers and charging duration of existing charging methods. Inductive charging with magnetic resonance coupling is being considered to offer a lot of positive marks over conductive charging being a more user handy, environment-friendly and a safer system. To set up a charging station, a well-conceptualized network is required. The basic need for this network is an adequately built charging infrastructure. These requirements include properly built stations having 24 × 7 hour power supply, economic installation and proper design of infrastructure. This paper outlines the challenges to the adoption of charging infrastructure for electric vehicles in the country and their potential solutions. Further, a detailed analysis of inductive (wireless) charging system for EVs in Indian context including the mathematical model and simulation results is outlined in this paper. An economic analysis pertaining to the deployment of wireless charging infrastructure in India has been done to evaluate the feasibility of the same in the present scenario. International standards for wireless charging are also taken into account while developing the hardware to address the safety, environmental and regulatory issues.

Keywords Inductive charging · Conductive charging · Magnetic resonance coupling · Charging infrastructure · Wireless charging · Charging duration

---

Vatsala · R. H. Khan · Y. Varshney · A. Ahmad · M. S. Alam (✉)
Centre of Advanced Research for Electrified Transportation,
Aligarh Muslim University, Aligarh, India
e-mail: hybridvehicle@gmail.com

R. C. Chaban
Hyundai Kia R&D, Detroit, MI, USA

© Springer Nature Singapore Pte Ltd. 2018
R. K. Pillai et al. (eds.), *ISGW 2017: Compendium of Technical Papers*, Lecture Notes in Electrical Engineering 487, https://doi.org/10.1007/978-981-10-8249-8_14

1 Introduction

The globe resources are being exploited at a very fast pace due to the rapid development of nations. The need of the hour is to have a safe and pollution-free environment. This can be done by reducing the excessive dependency on the available non-renewable resources which would be exhausted by some decades if used continuously. Numerous alternatives are being adapted to reduce this dependency, and one of the major steps taken in this direction is the electrification of gasoline engine vehicles, that run on battery. Electric vehicles are expected to revolutionize the global automotive industry in the upcoming years. Electric vehicles have a pre-installed battery that can be charged through various ways. The different methods to charge an EV are conductive charging and inductive charging also known as wireless charging. Conductive charging is done through wires that has three different levels of power transfer. Wireless charging can be deployed efficiently by inductive power transfer or capacitive power transfer.

Wireless charging comes with various advantages like safety, convenient charging in state of rest or motion of vehicle etc [1]. The charging of EVs is one of the main aspects to be considered for its successful administration among the general public which requires a public charging infrastructure.

The chief problems that need to be looked into are charging time, range of an EV and availability of electricity to provide 24×7 hour charging. The driving range of an EV is less and the plug-in charging takes a lot of time when compared with the refuelling of a gasoline engine vehicle which takes around 2 min [2]. To remove the constraints of charging time and range of an EV, wireless charging can be adopted for constant charging of vehicle on the route.

Wireless charging enables the automated charging of vehicles. It can be done in three ways, namely static, quasi-static and dynamic. The static charging system can be installed at home garages and parking lots, and EVs can be charged at any parking area reducing the driver's effort to connect the charger to vehicle and eliminating the shock hazards due to wires [3]. The quasi-static wireless charging system may provide energy to the vehicles not in motion but stopped for short duration such as on traffic lights, enhancing the vehicle range and decreasing the energy storage requirements in vehicles [4]. Dynamic wireless charging system provides continuous charging of vehicle while the vehicle is moving on a specified charging lane on the road, increasing the driving range and reducing battery size for xEVs [5]. Wireless power transfer with level 2 (230 V AC) charging at the rate of 7.2 kW has been achieved via wireless charging systems with an efficiency up to 88.5% [6]. Charging infrastructure for wireless charging systems will reduce the dependency on battery system with a minimal energy storage requirement only and convenience in charging. Major technical issues related to the charging infrastructure are its design, construction, operation and maintenance. Installation of charging infrastructure at short distances would deliberately reduce the vehicle cost and support an unlimited range for an xEV. The Government of India has taken an initiative under the name faster adoption and manufacturing of hybrid and electric vehicles in India (FAME India) which is a part of the National Electric Mobility Mission Plan which aims in ensuring a vehicle population of about 6–7 million electric/hybrid vehicles in India by the year 2020 [7]. The implementation of such a big scheme altogether requires a firm charging infrastructure to ensure India's global leadership in hybrid/electric vehicle segments.

2 Wireless Charging System

The main components of wireless charging system are AC–DC converter, DC–AC high-frequency inverter on the transmitting side, parking pad (consists of transmitter coil), vehicle adapter (consists of receiver coil) installed on the chassis of vehicle, an AC–DC rectification unit which can be on-board (i.e. vehicle integrated) or integrated in the vehicle adapter to match the power requirements for battery charging. The compensation circuits are used for impedance matching to get maximum power output for battery charging. When a vehicle is parked in the parking block, the receiver coil gets coupled magnetically with the transmitting coil and power is transferred via high-frequency resonant electromagnetic field set-up between the coils. A block diagram showing the working of static charging system is shown in Fig. 1.

2.1 Text Modelling of Wireless Charging System

Equivalent circuit of magnetically coupled resonant static wireless charging model is shown in Fig. 2.

Here the subscripts "1" and "2" denote the "transmitter" and "receiver" coil values of inductor, L, resistance, R, and capacitance, C. V_s is the source voltage of the primary circuit, equal to V_1, while R_l means the load resistance of the secondary circuit. When the operating angular frequency of the power source is ω, the mutual inductance M can be determined as follows (1):

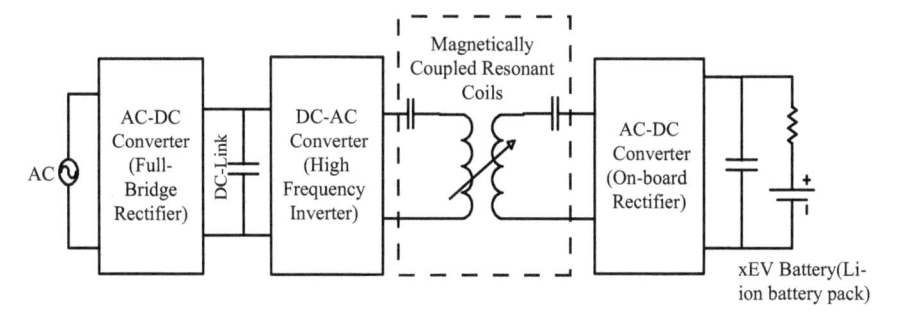

Fig. 1 xEV static wireless charging model

(a) Primary Series LC Compensation   Secondary Series LC Compensation

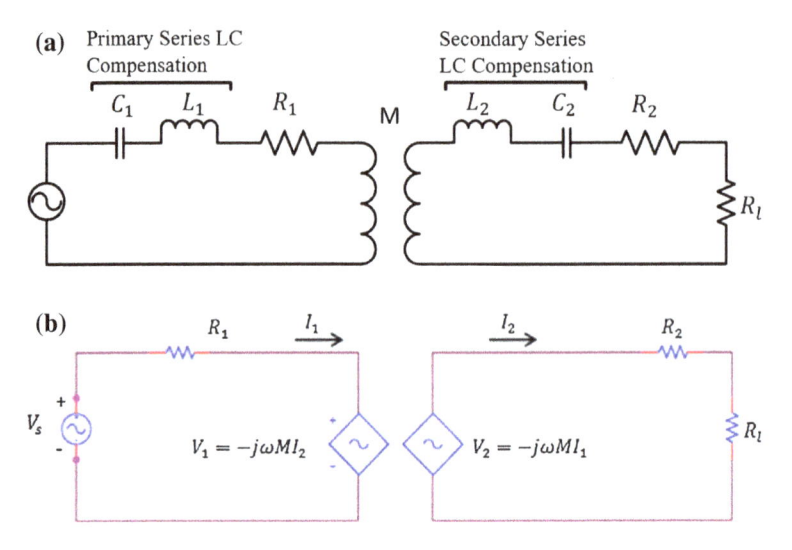

Fig. 2 Equivalent circuit of magnetically coupled transmitter and receiver coil. **a** Equivalent electrical circuit. **b** Equivalent circuit at resonance [8]

$$M = \frac{V_s}{\omega I_1} \qquad (1)$$

With the magnetic coupling of the two coils having coupling coefficient, k, the mutual inductance, M, can be expressed as below in (2).

$$M = k\sqrt{L_1 L_2} \qquad (2)$$

Now, the circuit in Fig. 2a can be modified to (b) by modelling the mutual inductance, M, in the equivalent circuit as in Fig. 2b.

$$V_1 = \left(\frac{1}{j\omega C_1} + j\omega L_1 + R_1\right) I_1 - j\omega M I_2 \qquad (3)$$

$$V_2 = j\omega M I_1 - \left(j\omega L_2 + \frac{1}{j\omega C_2} + R_t\right) I_2 \qquad (4)$$

Here, $R_t = (R_1 + R_2)$ represents the total resistance of the secondary circuit. Rearranging (3) and (4), we can get the expression for the current in the second circuit as in (5).

$$I_2 = \frac{j\omega M}{\left(R_1 + \frac{1}{j\omega C_1} + j\omega L_1\right) \times \left(R_t + \frac{1}{j\omega C_2} + j\omega L_2\right) + \omega^2 M^2} V_1 \tag{5}$$

From (5), the transfer function between the circuits in terms of the power supply voltage of the primary circuit and the output-induced current of the secondary circuit can be defined. At resonance, the criteria the capacitances in primary and secondary circuits should meet are expressed in (6) and (7), as shown in Fig. 2b, at the resonant frequency ω_0.

$$\frac{1}{j\omega_0 C_1} + j\omega_0 L_1 = 0 \tag{6}$$

$$\frac{1}{j\omega_0 C_2} + j\omega_0 L_2 = 0 \tag{7}$$

Applying the two Eqs. (6) and (7), both the circuits can be tuned for stronger magnetic coupling. At resonant frequency ω_0,

$$V_1 = R_1 I_1 - j\omega_0 M I_2 \tag{8}$$

$$V_2 = j\omega_0 M I_1 - R_t I_2 \tag{9}$$

Solving these mathematical equations, the transfer efficiency of the system comes out as (10).

$$\eta = \frac{1}{1 + \frac{R_1 R_t}{\omega_0^2 M^2}} \tag{10}$$

In terms of Q-factors of transmitter and receiver coils, the efficiency can be written as (11):

$$\eta = \frac{k^2 \cdot Q_1 \cdot Q_2}{1 + k^2 \cdot Q_1 \cdot Q_2} \tag{11}$$

where $Q_1 = \frac{\omega_0 L_1}{R_1}$, and $Q_2 = \frac{\omega_0 L_2}{R_2}$.

In order to have maximum power transfer (i.e. high efficiency) from primary to secondary circuit, the condition obtained from (10) and (11) is

$$\frac{R_1 R_t}{\omega_0^2 M^2} \ll 1 \tag{12}$$

$$k^2 \cdot Q_1 \cdot Q_2 \gg 1 \tag{13}$$

From (12), it can be realized that to obtain a highly efficient system, the operating resonant frequency along with the mutual inductance between the two resonating coils should have higher values. In other words, from (13), the Q-factors of the two coils and coupling coefficient between the two coils should be high.

3 Feasibility Analysis of Wireless Charging Model

A model to observe the magnetic field variation due to transmitter coil on receiver coil plane as shown in Fig. 3 is simulated in Ansys Maxwell 15.0.

Magnetic field is maximum at the central axis of transmitter coil, i.e. when both the coils are perfectly aligned. The magnetic field intensity decreases off the central axis, and thereby the value of coupling coefficient decreases, resulting in lower power transfer efficiency. Offset of receiver coil up to 5 cm is tolerable with effective efficiency of more than 80%.

4 Wireless Charging Standards

To develop safe, efficient and electromagnetic compatibility (EMC) wireless charging systems, different standards have been proposed. Two prominent standards for low-power WPT system which are supported by phone manufacturers are as follows: Q_i wireless power consortium (WPC) and alliance for wireless power (A4WP) [9]. The international standard for high-power WPT systems (3–20 kW), i.e. for xEVs is SAE J2954 which is industry-wide accepted standard for electric vehicles that outlines the minimum criteria for interoperability, minimum

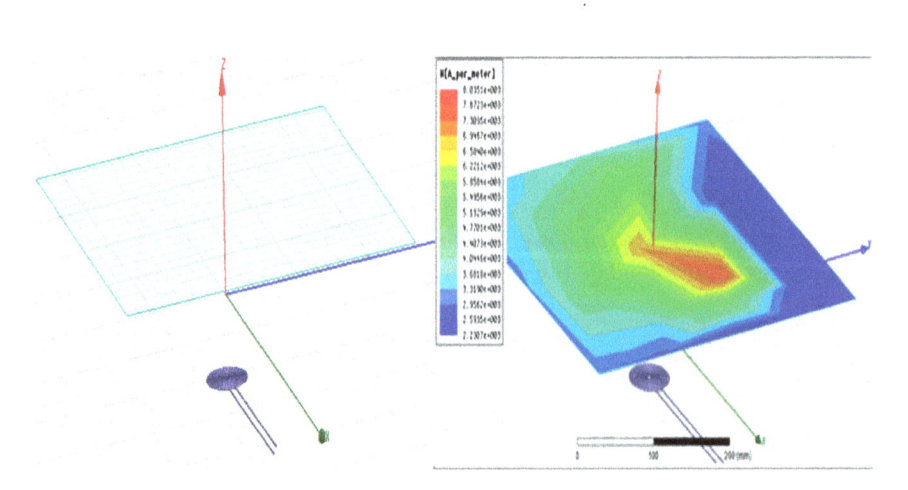

Fig. 3 Analysis report

performance and testing for wireless power transfer of electric and plug-in electric vehicles. It defines the WPT frequency band, positioning of vehicle and coil geometry, safety limit for electromagnetic field and wireless communication protocols. Some major auto OEMs like BMW, Honda, Toyota and Chrysler are developing vehicles under this standard.

The second main standard employed for fast DC charging is CHAdeMO which is supported by fast dc chargers for battery electric vehicles. CHAdeMO can charge low-range electric cars (120 km) in less than half hour with power level up to 62.5 kW. Also, it has got the recognition of IEEE STANDARD 2030.1.1TM-2015 in September 2015 prior to which it was already an IEC/EN standard [10]. ABB manufactures 50 kW (500 V, 100 A) and 20 kW CHAdeMO models with UL certification for the Americas markets.

5 Wireless Charging Infrastructure Model

In static charging system, the transmitter charging pads are fitted in the ground at the specified parking blocks. When a vehicle comes over the charging pad, the receiver coil on the vehicle gets coupled with the transmitter coil in the charging pad. After coupling, the vehicle is checked by its registration number, and if authorized, the transmitter coil is activated and charging begins. Each parking block has an installed charging pad with 24 × 7 hour electricity supply to provide any time charging facility for the vehicles. An animated model of a typical wireless charging is shown below in Fig. 4.

Fig. 4 Wireless charging infrastructure model

6 Challenges and Potential Solutions

Though there are plentiful merits to the concept of wireless charging infrastructure, they are equally backed up with a handful of challenges relating to its design, construction, operation and maintenance [11]. Definitely, the advantages cover the demerits, but the demerits cannot be neglected. Installations of private charging centres serving domestic vehicle charging meant for 2–3 vehicles prove to be very costly. The equipment used in power transfer is expensive and cannot be afforded to replace frequently.

Here, we enumerate the various challenges to set up wireless charging stations in India as in the present scenario.

6.1 Power Handling Challenges

Challenges related to power handling can be categorized as follows:

- Smooth power installation systems.
- Increase in power demands from grids.
- Its effect on other electricity-hungry industries.

Intake of energy from the renewable energy sources such as solar and wind power plants would safeguard the grid from overloading. Integration of the charging infrastructure with the smart micro-grids would add new feature to the charging infrastructure; i.e. vehicles can give back power to the grid through vehicle to grid (V2G) application when ample amount of energy is stored in batteries [12]. A proper charging infrastructure needs an efficient power installation system that includes power conditioning units which changes the frequency on which the system works. In order to set up an infrastructure of such kind, a load must be dedicated completely for the purpose.

6.2 Deployment of Optimal Charging Infrastructure

An xEV typically has a range of 150 miles in one run which is very less as compared to the gasoline vehicles. In order to fill up a gasoline vehicle, it takes 3 min at the cost of some Rs. 2000 in India, whereas to recharge a 25 kWh electric vehicle, it takes a few hours but a much neglected cost of Rs. 250 in the same country. The cost has been reduced tremendously on account of increased refuelling time.

In order to come up with driving range and charging time issues, charging points may be deployed at short distances which would also reduce the energy storage capacity required for xEVs, thus reducing the cost and size of battery, resulting in lowering of an xEV cost. This solution involves a huge amount of investment which can be recovered by charging the drivers with the appropriate costs.

6.3 Efficiency Management

Though the Clemson University's International Centre for Automotive Research (CU-ICAR) along with the International Transportation Innovation Centre (ITIC) recently came out with a static wireless power transfer technology having a 6.9 kW rate and an average efficiency of 85% [13], yet improved schemes capable of getting commercialized need to be devised to make a system efficient as it would lead to smaller charging times.

6.4 Maintenance Issues

Maintenance of such an infrastructure becomes very important, being a costly investment. This role if left unattended may lead to a considerable loss. The major issues are as listed below:

- Proper equipment handling.
- Wear and tear issues related to machinery as well as the construction.
- Timely replacement of worn-out parts.
- Periodic checks and assessments.

For India, being a densely populated country, it is difficult to acquire large land areas at short distance solely for charging of vehicles. To seek an effective solution, parking lots of shopping malls, hotels, large corporate offices, apartments, home garages and other public parking spaces can serve for the purpose. Multi-storied set-ups can also be constructed to eradicate spacing issues in metropolitans.

7 Economic Analysis

To check the commercial feasibility of the wireless charging system, an economic analysis has been done in view of different public utilities parking area. The cost of charging is calculated for the xEVs based on a payback period of 5 years of the wireless charging system cost and annual commercial electricity costs. Some assumptions taken while doing the study are as follows:

(1) Installation and land costs are neglected.
(2) The parking is assumed to be fully equipped for wireless charging in case of hotels and residential complexes considering these to give excellent guest services.
(3) The present costs of electricity charges are considered, and the source of electricity is considered to be thermal power stations.
(4) The cost of petrol/diesel is unstable depending upon various factors including government regulations.

7.1 Cost Estimation of Charging System

The cost of level 2 (230 V AC) wireless charging system with power rating 3.3 kW is estimated, and break-up cost of different components of the system is shown in Table 1 (Fig. 5).

For transmitter coil of charging pad, the transmitter coil is considered to be of 12 AWG wire and the specifications are given in Tables 2 and 3.

$$\text{Length of spiral} = \frac{N^2 * A^2}{30A - 11D_i} \quad (14)$$

where N = Number of turns in coil

$$A = \frac{D_i + N(W + S)}{2} \quad (15)$$

where

W Wire Diameter,
S Turn Spacing and
D_i Inner diameter of spiral (in mm).

Table 1 Cost of single charging system

Equipment	Costs (₹)
AC–DC converter	2720–3400
High-frequency inverter	8160–11,000
Transmitter coil mechanism	14,212
Total cost	25,840

Fig. 5 Cash flow for maintenance and initial costs incurred

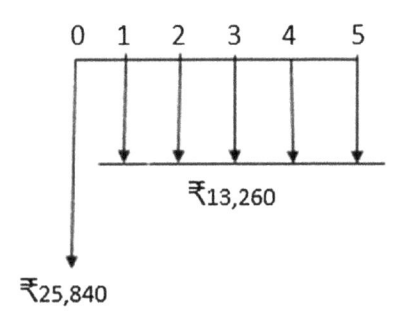

Table 2 Specifications of transmitter spiral coil

Coil parameters	Dimensions (cm)
Outer diameter	30
Diameter of wire	0.205
Turn spacing	0.1
Length of wire	2542

Technical and Economic Feasibility Analysis for deployment of xEV ...

Table 3 Set-up of wireless charging system for public conveniences

Public convenience	Data obtained from	Car capacity	No. of charging systems required (per hour basis)	Total cost of charging system to be installed (₹)
Airport	IGI airport, New Delhi	4300	2200	56,848,000
Hospital	Hrudalaya, Bengaluru	400	200	5,168,000
Hotel	ITC Grand Chola, Chennai	500	500	12,920,000
Metro station	DMRC, New Delhi	450	350	9,044,000
Residential complex	RNA regency park, Mumbai	75	75	1,938,000
Shopping complex	Lulu mall, Ernakulam	3000	1500	38,760,000
Exhibition centre	India Expo Mart, Noida	4000	2000	51,680,000

Commercial cost of procuring electricity = ₹10/Unit (as per Indian costs).
Initial installation cost = ₹25,840 from Table 1.
Estimated annual maintenance cost = ₹13,260.
The inflation rate will be around 4.80% in 2020 according to present economic trends. The interest rate is around 10.50% for commercial purchase (both compounded annually). Therefore, calculating the annuity

$$A = P\left[\frac{i_{\text{eff}}(1+i_{\text{eff}})^n}{(1+i_{\text{eff}})^n - 1}\right] \quad (16)$$

where

$$i_{\text{eff}} = (1+i)(1+f) - 1 \quad (17)$$

The i_{eff} given in eq.(17) is the effective interest rate obtained after incorporating MARR (i) and inflation rate (f). Therefore, $i_{\text{eff}} = 15.80\%$. We consider the system to be deployed for 12 h/day for 365 days. The cost evaluation for per hour charging has been calculated in Table 4.

Table 4 Cost evaluation to evaluate cost of per hour charging

Costs	Rupees (₹)
Annual installation recovery	7854
Annual maintenance recovery	13,260
Commercial electricity costs	144,540
Total annual return needed	165,597
Cost of per hour charging	37.8

7.2 Comparison of Electric Vehicle with Gasoline Vehicle

We consider the two cars FORD FIGO and NISSAN LEAF from the midsize segment of FORD and NISSAN. The average mileage for FORD FIGO (petrol) is 18 kmpl, and for FORD FIGO (diesel) is 25 kmpl [14]. The fuel tank capacity is 42 l for FORD FIGO. The prevailing rate of diesel (New Delhi) is ₹55, and the prevailing rate of petrol (New Delhi) is ₹64. The range per 100% charge is 135 km for NISSAN LEAF. The time required to charge the battery is about 7 h [15]. The cost of per hour charging is calculated to be ₹40 (approximately) (as shown in Table 4). The cost per km for EVs and gasoline vehicles has been evaluated and compared as shown in Table 5, and therefore, Fig. 6 shows the comparison between the gasoline and electric vehicles on the basis of fuel expenditure.

The environmental analysis for wirelessly charged EVs and gasoline vehicles is shown in Table 6 and Fig. 6.

From Fig. 6, we observe that for the same range, wirelessly charged electric vehicles have the advantage of lower fuel expenditure. Further, from Table 6 and Figs. 7 and 8, it is observed that the emission levels by ICEV are comparatively higher than EVs.

Table 5 Comparison of an EV with gasoline vehicle

Type of vehicle	Range with full fuel (km)	Total cost (₹)	Cost per km (₹)
FIGO (Petrol)	756	2688	3.5
FIGO (Diesel)	1050	2310	2.2
NISSAN LEAF	135	280	2.07

Fig. 6 Comparison of gasoline vehicles with respect to electric vehicles (for range = 10,000 km)

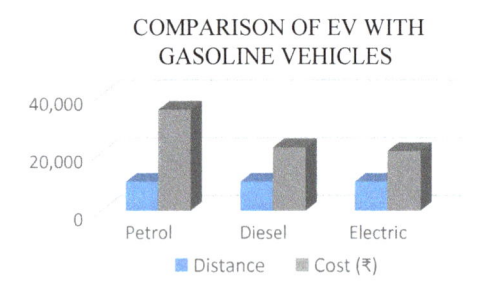

Table 6 Emission comparison for electric vehicle and ICEV

Pollutant (g/km)	Electric vehicle	ICE vehicle
CO_2	130	139
NOx	0.05	0.0255
CH	0.055	0.2
PM	0.045	0.075

Fig. 7 Comparison of CO_2 (a major greenhouse pollutant) emissions by ICEV and EV

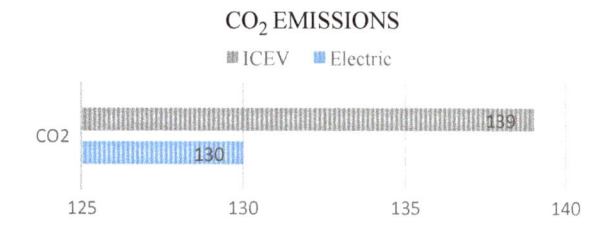

Fig. 8 Comparison between environmental emission by ICEV and EV

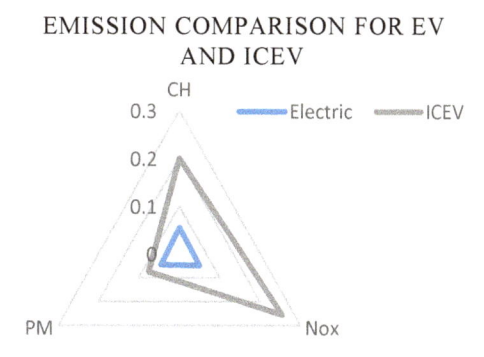

8 Conclusions

This paper shows the need of inclination towards electrified transportation. Various challenges to the installation of wireless charging systems for EVs are discussed along with their potential solutions. A detailed economic analysis has been presented, and different public conveniences have been surveyed. The results clearly depict that the wirelessly charged EVs not only provide convenience to the user but also help to conserve the fossil fuels ultimately leading to a less polluted environment at lower operating costs.

References

1. Suh I-S, Kim J-D (2014) Design considerations of wireless power transfer applications to electric vehicle charging in efficiency and misalignment. In: Proceedings of Asia-Pacific microwave conference
2. Yilmaz M, Krein PT (2013 May) Review of battery charger topologies, charging power levels, and infrastructure for plug-in electric and hybrid vehicles. IEEE Trans Power Electron 28:2151–2169
3. Li S, Mi CC (2015 Mar) Wireless power transfer for electric vehicle applications. IEEE J Emerg Sel Top Power Electron 3:4–17
4. Wang L, Gonder J, Burton E, Brooker A, Meintz A, Arnaud (2015 Nov) A cost effectiveness analysis of quasi-static wireless power transfer for plug-in hybrid electric transit buses. IEEE–Veh Power Propul Conf (NREL/CP-5400-64089)

5. Lukic S, Pantic Z (2013 Oct) Cutting the cord: static and dynamic inductive wireless charging of electric vehicles. IEEE Electrification Mag 1:57–64
6. Zhu Q, Wang L, Wang L (2014) Compensate capacitor optimization for kilowatt-level magnetically resonant wireless charging system. IEEE Trans Industr Electron 61:6758–6768
7. National Electric Mobility Mission Plan (NEMMP-2020) [Online] Available http://dhi.nic.in/UserView/index?mid=1347
8. Suh I-S, Kim J (2013) Electric vehicle on-road dynamic charging system with wireless power transfer technology. In: IEEE international electric machines and drives conference (IEMDC)
9. Lu X, Wang P, Niyato D, Kim DI, Han Z (2015 Nov) Wireless charging technologies: fundamentals, standards, and network applications. IEEE Commun Surv Tutorials 18:1413–1452
10. Technical Specifications of a DC Quick Charger for Use with Electric Vehicles, IEEE Std. 2030.1.1-2015
11. Gill JS, Bhasvar P, Chowdhury M, Jhonson J, Taiber J, Fries R (2014) Infrastructure cost issues related to inductively power transfer for electric vehicles. International conference on ambient systems, networks and technologies (ANT-2014)
12. Nooria M, Zhaoa Y, Onatb NC, Gardnerc S, Tataria O (2016) Light-duty electric vehicles to improve the integrity of the electricity grid through vehicle-to-grid technology: analysis of regional net revenue and emissions savings. Elsevier Appl Energy 168:146–158
13. CU-ICAR, SCTAC partner on wireless vehicle charging [Online] (2013 July) Available http://newsstand.clemson.edu/mediarelations/cu-icar-sctac-partner-on-wireless-vehicle-charging/
14. Figo Mileage-ARAI [Online] Available https://www.cartrade.com/ford-cars/figo/mileage
15. NISSAN LEAF RANGING AND CHARGING [Online] Available https://www.nissan.co.uk/vehicles/new-vehicles/leaf/charging-range

Viability of xEVs in India: A Public Opinion Survey

Mohammad Asaad, Prashant Shrivastava, Mohammad Saad Alam, Yasser Rafat and Reji Kumar Pillai

Abstract Electric vehicles (EVs) are an important aspect of the ongoing bid for the cleaner mode of transportation as the critical issues like climate change, energy security, and declining energy reserves direly need attention. Like any new technology, their acceptance into society has been a dull affair. Society has a tendency to show reluctance in accepting a new form of technology that is yet to be matured. As Government of India, under FAME mission, plans to roll out about 6–7 million electric/hybrid vehicles by the year 2020, a positive consumer mindset in accepting EVs is crucial for their success. A questionnaire-based survey was conducted and response of people from various domains, viz. service, R & D, industry, academia was recorded. Based on their responses, this paper provides valuable insights into consumers' attitude toward attractive features of EVs, acceptable range, disadvantages, key roadblocks in reception and consumers' willingness to pay extra compared to conventional vehicles.

Keywords Electric vehicles · Indian scenario · Consumer attitude
Introduction barriers

M. Asaad · P. Shrivastava · M. S. Alam (✉) · Y. Rafat
Centre of Advanced Research in Electrified Transportation,
Aligarh Muslim University, Aligarh, India
e-mail: saad.alam@zhcet.ac.in

M. Asaad
e-mail: mohammad.asaad@zhcet.ac.in

P. Shrivastava
e-mail: prashant.xev.ess@gmail.com

Y. Rafat
e-mail: yasser.rafat@zhcet.ac.in

R. K. Pillai
India Smart Grid Forum (ISGF), New Delhi, India
e-mail: reji@rejikumar.com

1 Introduction

1.1 Background

With the ongoing endeavor for an emission-free environment, the transportation sector is a large contributor to this cause. This sector produces a large quantity of CO_2 and other harmful gases like sulfur oxides (SOx), nitrous oxides (NOx), unburnt hydrocarbons. The global greenhouse gas emissions of transportation sector account for 14% of total emissions and are predicted to rise to 50% by the year 2030 [1]. This prediction indicates that the existing transportation system is not viable for sustaining a clean atmosphere.

The deterioration of air quality, issues of global warming, and continuous depletion of fossil fuels are serious issues to consider. The advent of electrification of transportation is aimed to abate the ever-continuing emission level and more importantly to cut economic dependence on oil and accordingly reduce detrimental dependence on petroleum exporting economies which largely exist in Middle East region. The importing of oil can become an issue as political and civil disputes in this region, viz. oil embargo in 1973 and 1977, the Gulf war, the Iran-Iraq war disrupt the oil supply [2].

The electric vehicle (EV) technology exists for more than a century peaking commercially around 1900 [3]. However, due to the easy availability of fossil fuels, maturity in technology, and simplicity in the use of combustion engines, EVs were put on hold and limited to golf carts and delivery vehicles [4]. The need for a green and sustainable mode of transportation and advancement of battery technology has again sparked interest in EVs as a viable mode of transportation.

India is one of the top automotive markets in the world today and has a very fast rising population with buying potential and solid economic progress [5]. The automotive sector contributes 7.1% to the national GDP [6].

This large no of vehicle also produces a huge amount of emissions. Eighty-seven percentage of the total GHGs emissions in India is credited to road vehicles. Figure 1 shows the per capita CO_2 emissions in India over the years [7].

Fig. 1 Per capita CO_2 emissions in India over the years

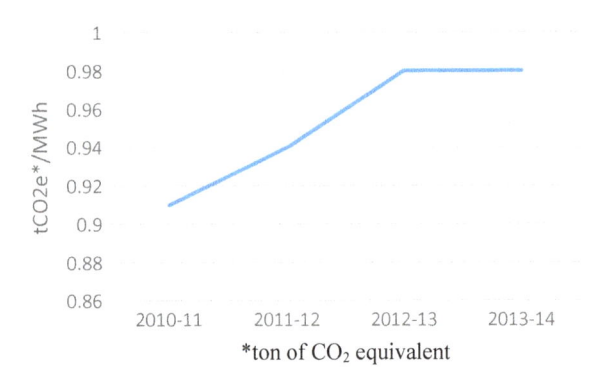

*ton of CO_2 equivalent

In India, air quality has reached an alarmingly high level. A large no of metropolitan cities have exceeded World Health Organization (WHO) standards [8].

If there is no reduction in present energy consumption style, India's primary energy consumption is expected to increase by 70% in the next ten years [9].

As India is an emerging economy, the dire need to find viable alternatives for sustainable transportation is emphasized by fast economic development so demand for transportation is fast-tracking, widening the gap between domestic fuel production and consumption. This along with an increase in global crude oil prices is expanding trade deficit which poses a threat to India's energy security [10].

The Government of India unveiled the National Electric Mobility Mission Plan 2020 (NEMMP) in the year 2013, which envisioned the sale of 6–7 million xEVs (full range of electric vehicles) in the Indian market by 2020. This plan would focus on four key areas—technology advancement, mandate creation, pilot projects, and charging infrastructure of xEVs [11]. The NEMMP 2020 is envisioned to provide the future roadmap, institute common set of principles, comprehensive norms, and framework for endorsing the adoption of the full range of electric mobility solutions for India, which can augment national fuel security, provide inexpensive and environmentally friendly transportation, and empower the Indian automotive industry to attain global manufacturing leadership. In April 2015, the government formulated a scheme for faster adoption and manufacturing of electric vehicles (FAME) in India, under the National Electric Mobility Mission Plan 2020 to encourage the progressive induction of reliable, affordable, and proficient xEVs. Accounting for a high level of emissions in high-density urban centers, this scheme will be restricted to smart cities, major metro cities, e.g., Delhi-NCR, Mumbai, Kolkata, Chennai, Bangalore, Ahmadabad, Hyderabad, all state capitals and cities with more than 1 million population and cities of northeastern states [12].

Common electric cars in today's Indian market are presented in Table 1.

xEVs exist in many types on the basis of the range of electrification of the drive train. Most popular types are hybrid electric vehicle (HEV) which use an internal combustion (IC) engine along with a battery operated electric motor, plug-in hybrid electric vehicle (PHEV) which are EVs with smaller IC engine and powerful electric storage, i.e., battery that can also be recharged, battery electric vehicle (BEV) is a purely electric vehicle and obtains motor power completely from battery that can be plugged into recharge through electric supply [13].

Table 1 Popular electric cars in India

Name	Manufacturer	Type
E2O	Mahindra Reva	BEV
XC90 T8	Volvo	PHEV
Camry	Toyota	HEV
i8	BMW	HEV
eVerito	Mahindra	HEV
eSupro	Mahindra	BEV

1.2 Need for This Survey

Electric vehicles are an emerging technology in Indian scenario so they require a change in the mentality of consumers to make them commercially viable, especially due to some key differences with their counterpart internal combustion engine (ICE) vehicles, viz. high initial cost and limited range.

In spite of xEV offering efficient and clean fuel, consumer's preference goes toward conventional vehicles (CVs).

High initial investment, low-risk tolerance, and lack of knowledge by prospective consumers are some common hurdles to the acceptance of any new technology [14].

Research by Oliver and Rosen [15] implies that market acceptance of xEVs is restricted partially due to apparent risks with innovative technologies and balance between fuel efficiency, sizing, and cost.

Past trend in acceptance of technology illustrates that while latest technology is innately appealing to a few primary adopters, including futurist and technology enthusiasts, the mainstream customer will continue to be skeptic about the new technology [16].

The common customer, when choosing, tends to like "notions of tradition and familiarity" instead of accepting a modern technology [17].

This study is aimed to study public's perception about electric vehicles and common issues of their reluctance in accepting it. Also, their feedback is taken on what they consider a major deterrent in purchasing an xEV.

2 Methodology

To achieve the research goals, a questionnaire-based survey method was chosen. This method allows collecting large and widespread data in a short period. The sample population consists mostly of current users of ICE vehicles with the objective of portraying their views and mindsets as they are likely to be future owners of xEVs.

The survey was conducted majorly at three events. First at a workshop "EV Boot Camp" organized by Center of Advanced Research in Electrified Transportation (CARET) at Aligarh Muslim University, Aligarh, India.

The second event is the Indian Smart Grid Forum (ISGF) week 2016 (New Delhi, India).

The third event is the Symposium on International Automotive Technology (SIAT) 2017 organized by Automotive Research Association of India (ARAI) at Pune, India.

Rest of the responses was taken in an interactive online-based format where 136 responses of various strata of the population were noted.

Figure 2 shows the region-wise composition of the sample taken.

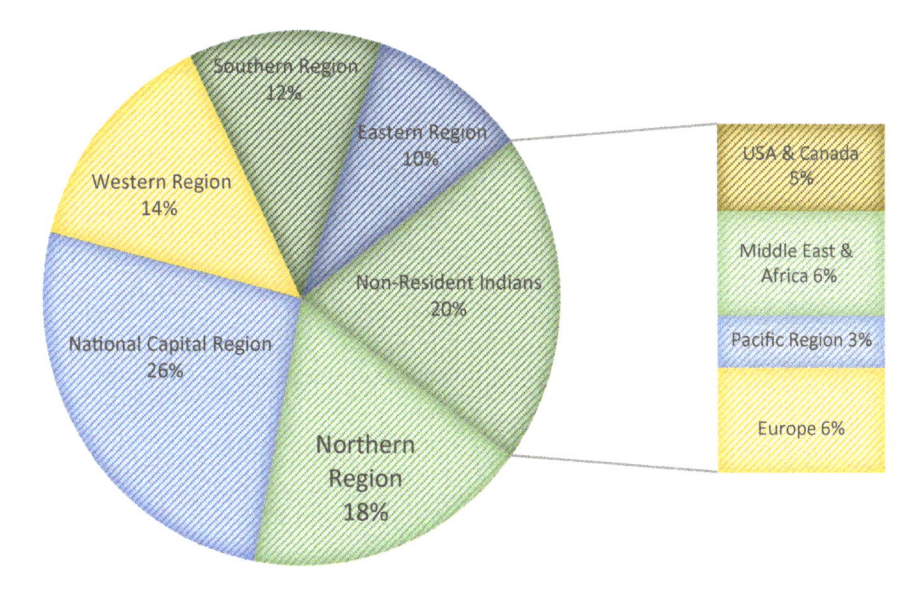

Fig. 2 Region-wise composition

In terms of technical awareness, the large proportion of the survey population has been considered as technology enthusiasts. These individuals are well versed in worldwide technology advancement, have high perception skills, and further prepared to classify the many technical, economic, and ecological dissimilarities between xEVs and ICE vehicles. For this survey, these individuals are considered to be probable primary adopters if they recognize xEVs to be better in comparison to conventional vehicles performance wise. It should be stated that the sample population may not essentially be demonstrative of the general populace; however, it presents useful data about technology enthusiasts. Table 2 represents the demographic composition of the population sample.

3 Survey Results and Discussion

3.1 Interest in xEVs

With hybrid electric vehicle (HEV) being more prevalent in market tailed by plug-in hybrid electric vehicle (PHEV) and then battery electric vehicle (BEV), the level of interest by respondents in different xEV types follows the same trend. As shown in Fig. 1, HEVs' (37%) appeal is fairly huge than pure electric BEV (28%) due to large range advantage for a lengthy commute.

Table 2 Demographic composition of the sample popular electric cars in India

Characteristics		%
Sample size		354
Gender	Male	62%
	Female	38%
Age	18–25	40%
	26–45	34%
	46 and over	26%
Occupation	Academia:	
	Students	16%
	Faculty	12%
	Automotive dealers/vendors	11%
	OEM	20%
	Automotive parts suppliers	10%
	Business class	09%
	Government. officials	07%
	R & D	10%
	Small and medium enterprise	05%

Advantages identified with xEVs by consumers were regarding environmental aspects, battery performance aspects, efficiency, reduced dependence on fossil fuels, and future energy security.

As xEVs reduce the cost per 100 km of driving (initial car price and maintenance cost neglected), the majority of respondents consider 50% reduction in per 100 km cost would be enough for them to switch to an xEV in India.

When asked about willingness to pay extra, respondents were ready to shell out 10% extra for a new xEV instead of new ICE vehicle.

In terms of various characteristics of xEVs, reduction in air pollution seems to be most appealing attribute to respondents followed by cutback or elimination of fossil fuels (Fig. 3). Luxury and looks/style received the lowest ranking.

Many respondents did not realize that all-electric cars require less upkeep than conventional cars, e.g., Oil/coolant change is not necessary plus fewer fragile parts. Also, people underestimate the fuel savings electric cars offer.

Lack of knowledge of attributes and features for comparison to assess their xEV choices severely limits the consumers' willingness to own an xEV and also, they

Fig. 3 Demand for different xEV types

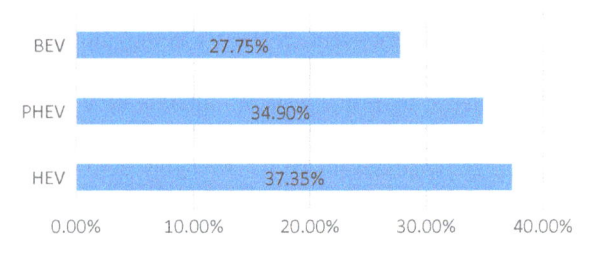

BEV 27.75%
PHEV 34.90%
HEV 37.35%

Viability of xEVs in India ... 171

are well acquainted with buying, driving, and fueling ICE vehicles. The well-developed required framework including dealerships, service stations, roadside assistance, and a huge network of petrol stations makes difficult to unseat conventional vehicles.

Furthermore, the presence of regenerative braking, which is new to ICE users, provides a rare feeling. When assimilated into the vehicular system, it slows the vehicle and captures energy the moment driver removes his foot from the accelerator pedal, thus being advantageous in heavy traffic.

3.2 Concerns About xEVs

As per responses, a major concern cited was the high initial cost (30%) closely followed by limited battery range (26%) and lack of charging infrastructure (25%). These concerns highlight the key drawbacks of xEVs when compared to ICE vehicles. A full breakdown of concerns is presented in Fig. 4.

The fact that only 4% of respondents identified safety as the major concern suggest that majority of technology enthusiasts consider xEV as a safe mode of transportation.

The many constant issues that influence consumers' indecision and skepticism about the judgment of an xEV purchase combined with the cost and range anxiety, adversely affect the EV sales. Insecurity and apparent risk affect buyer's inclination to buy modern inventions, especially costly, long-lasting ones, e.g., automobiles.

Fig. 4 Key characteristics of xEVs

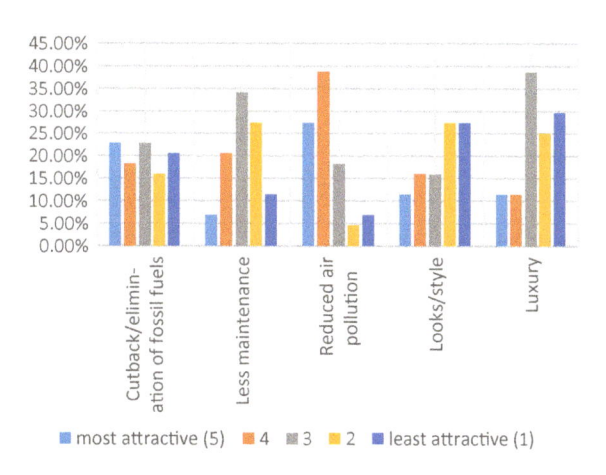

3.3 Driving Range and Battery Charging

Even though only a small number of Indian citizens travel large distance per day by car, consumers perceive the limitation of around 100–200 km as a struggle, which is the current range of EVs. This distance roughly equals the number of kilometers one can drive when the fuel gauge hovers over the red area (i.e., low fuel warning) in a conventional vehicle.

However, for occasional lengthy road trips, batteries need to be recharged during the trip. Also, long-range capability is a very desirable feature but with the range of the battery, the cost also increases. The question is: what is the minimum range that consumer requires for considering the purchase of an EV?

When asked about an acceptable driving range of EVs, general consumer opinion was 250 km.

Use of fast chargers is another option for long road trips with EVs. Fast chargers refer to quick charging (typically 30–45 min) of EVs at higher power levels. Around 60% of consumers were willing to add 30–45 min to their trip as a consequence to pull over every 200 km to fast charge.

There is one more choice for a long trip with EVs, that is, the battery swapping. It means the swift replacement of an EVs drained battery with a completely charged one at a battery switch station. In such case, the battery proprietorship is likely to be unconnected with vehicle proprietorship, i.e., the initial investment on EVs would reduce but a monthly subscription cost will be added just like an internet plan to cover battery proprietorship cost and the expenses to recharge and/or swap the batteries [13]. A large no of the respondent (63%) showed a willingness to buy an EV if they could purchase it without battery pack and then lease the batteries from the battery swapping station, on assumption that many battery swapping stations are around.

If given a choice between fast charging and battery swapping, general preference was toward fast chargers due to its cost effectiveness, even though it takes longer than battery swap (Fig. 5).

Fig. 5 Major concerns about xEVs

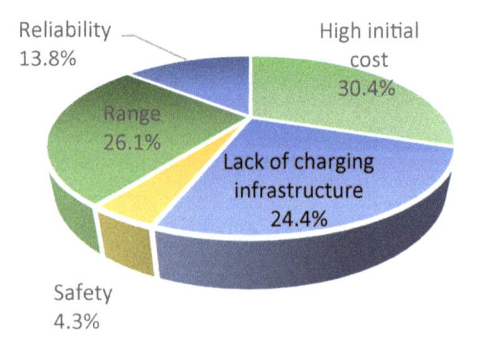

Charging of EVs can be done at many different locations, viz. home/parking lot of an office or apartment or public charging infrastructure. According to this survey, most preferred location is at home/parking lot of an office or apartment (62%).

3.4 Indian Scenario

Despite many advantages, acceptance of xEVs still faces major difficulties in extensive market reach [18]. Imply that battery technology restrictions and high battery price are the main hindrances in the widespread acceptance of xEVs.

To achieve a large-scale commercial success, xEVs need to overcome socioeconomic issues related to consumers. Adoption by the consumer is significant to the steady success of a viable transportation sector [19].

This survey opinionated consumers on prospects of xEVs in Indian scenario and majority (67%) considered them an integral part of transportation sector but only next to ICE vehicles. Figure 6 shows general opinion in the case of India (Fig. 7).

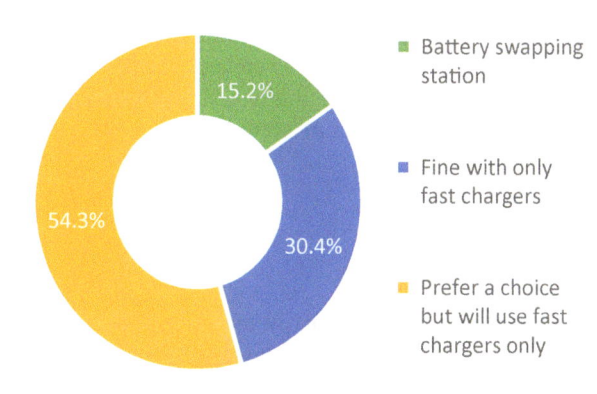

Fig. 6 Preferred method of charging on long trip

Fig. 7 General opinion on xEVs scenario in India

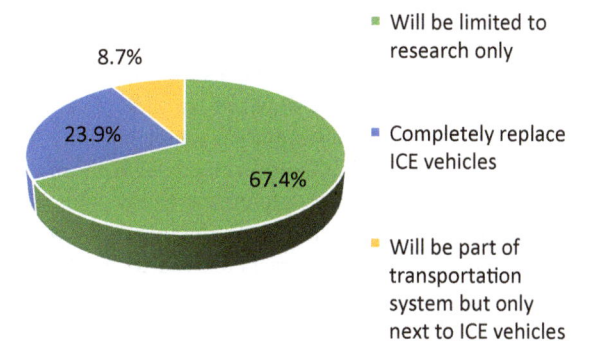

4 Potential Solutions to Major Public Concerns

Given the complexity of the concerns consumers face in the xEV purchase decision, several reliable, and concise information sources, e.g., advertising of vehicle manufacturer, internet resources should be made available for consumers for decision making. As public attitude can be impelled by media and social networks, policymakers can use this mode to sway the society admiration for non-financial advantages of adopting xEVs such as energy security and elimination/cutback of emissions.

Also, vehicle dealerships along with electric utilities can work to endorse the vehicles by accentuating the suitability and affordability of electric fuel.

The test-drive experience is a significant feature of the user assessment of xEV purchase as it involves a chance to get accustomed to EV's range and charging. Thus, more programs that offer "ride and drives" for xEVs at particular spot can be effective. The government should also explore opportunities for automotive industry players to offer a local EV know-how facility to offer essential test-drive prospects.

To instill confidence in masses of government's initiative and to enhance the visibility of xEVs generally, the government should exhibit leadership by adding xEVs to its fleets and establishing charging infrastructure at its facilities.

Other measures include awareness, bigger investments in xEV technology, infrastructure, battery swap stations, strong warranties on the xEV batteries, and tax waiver to reduce the cost of xEVs.

5 Conclusion

The sample used in this survey may not be the representative of the entire population due to differences in environmental awareness; however, it presents a useful understanding of inclination and mindset of various tech-savvy people.

Today foremost challenges faced by xEVs comprise of limitation of battery technology, high battery price, and lack of charging infrastructure. Also, user reception is imperative as it is fundamental to the commercial success (or failure) of xEVs, even if all the key challenges are overcome.

The result of the survey shows about 67% of consumers consider future of xEVs as an integral part of transportation sector but only in backstage to ICE vehicles.

Keeping in trend with the prevalence of HEV in the market followed by PHEV and then BEV, consumers' preference follows the same suit.

The results also show that the reduction in air pollution followed by cutback or elimination of fossil fuels are perceived to be the most appealing feature of the xEVs, while, high initial cost, limited driving range, lack of charging infrastructure are the key drawbacks of EVs. For the majority of the respondents, a range of 250 km is acceptable.

In the case of longer road trips, about 60% respondents gave preference to adding extra 30–45 min to trip in order to supercharge instead of swapping out batteries as supercharging is more cost effective. The vast majority of respondents preferred home/parking lot of an office or apartment for charging their xEV compared to public charging infrastructure.

Potential measures include advertisement in digital or print media to influence public opinion, education on affordability and sustainability of electric fuel, test-drive programs to get society accustomed to xEV experience, induction of xEVs into government/corporate fleets, bigger investments in xEV technology, infrastructure, battery swap stations, strong warranties on the xEV batteries, and tax waiver to reduce the cost of xEVs.

Acknowledgements This research is supported by Centre of Advanced Research in Electrified Transportation (CARET), Aligarh, India. We wish to acknowledge organizers and participants of EV Boot Camp 2016, Aligarh Muslim University, Aligarh, India, for permitting and participating in this study. We are grateful to delegates at India Smart Grid Forum (ISGF) week 2016 for providing their valuable feedback. We are also thankful to delegates at SIAT 2017 held in Pune, India, for sharing their keen insights.

Appendix

Survey questionnaire-
Name: **Gender:**
Designation:
Email:
Residential State:

1. Type, brand, and model of your current vehicle:

2. What is your general opinion on the future of all types of electric vehicles (xEVs) in Indian scenario?

(a) They will completely replace gasoline-powered cars in the following years
(b) They will be a part of the transportation system but will never take the throne from the gasoline-powered vehicles
(c) They will always be limited to research and will remain beyond the reach of masses.

3. Assuming xEVs reduce the cost per 100 km of driving (initial car price and maintenance cost neglected), how much reduction would be enough for you to switch to an xEV in India?

(a) up to 10%
(b) up to 25%
(c) up to 50%
(d) up to 100%

4. **How much more would you be willing to pay for a new xEV instead of a new gasoline-powered vehicle?**

(a) up to 10%
(b) up to 25%
(c) up to 50%
(d) up to 100%

5. **What according to you is the main reason for people not buying an xEV in India?**

(a) Not interested in a new car at all
(b) Expensive when paralleled to conventional internal combustion cars
(c) Lack of charging infrastructure
(d) Lower range with fully charged EV available as compared to fully tanked gasoline vehicle
(e) Reason not mentioned above

6. **In case you owned an xEV, where will you be charging it?**

(a) Home/Parking lot of an office or apartment
(b) Public charging infrastructure

7. **In case you went on a longer trip with an EV, would you consider using only fast chargers, i.e., pulling over for half an hour every 200 km in order to recharge at nominal cost, or would you insist on using a battery swapping station as a mean for receiving a full charge within a minute and pay higher tariff?**

(a) I am perfectly fine with only fast chargers
(b) I would like to have a choice, but I would always use a fast charger as it is cost effective
(c) I would definitely use a battery swapping station whenever I can, even if it is costlier than the fast charger.

8. **Would you be willing to add half an hour to your trip as a consequence of having to pull over every 200 km in order to fast charge?**

(a) Yes
(b) No

9. **Would you be more likely to buy an xEV if you could purchase it without battery pack (i.e., the investment reduction by 50%) and then lease the batteries from the battery swapping station? Assume that there are many battery swapping stations around**

(a) Yes
(b) No

10. Give rank to the following xEVs in terms of which interests you the most (1 being the least interesting and 3 being the most interesting)

(a) An internal combustion engine (ICE) is an engine used in most conventional cars in which combustion of fuel (usually gas and petrol/diesel) occurs
(b) A hybrid electric vehicle (HEV) is a combination of both electric vehicle and conventional vehicle (IC engine vehicle)
(c) A plug-in hybrid electric vehicle (PHEV) is a HEV but its battery can be recharged via the electric grid, providing purely electric power for a limited range
(d) A battery electric vehicle (BEV) operates solely on an electric battery and also features a plug-in charger

11. As the size of an xEV battery increases, the range increases, but so does the cost. With that in mind, how many kilometers minimum would the vehicle range have to be before you would consider buying a battery electric vehicle (BEV):

12. What do you consider your biggest concern about xEVs?

(a) High cost
(b) Battery range
(c) Safety
(d) Reliability
(e) Charging infrastructure
(f) Other, (please specify).

13. Please rank the following attributes of xEVs in terms of which appeals to you the most (1 being the least appealing and 5 being the most appealing)

(a) Decrease/eliminate the use of petroleum
(b) Less maintenance
(c) Reduced air pollution
(d) Looks/style
(c) Comfort

References

1. Edenhofer O, et al (2014) Climate change 2014: mitigation of climate change
2. Ehsani M, Gao Y, Emadi A (2010) Modern electric, hybrid electric and fuel cell vehicles, 2nd edn.
3. Matulka R (2014) The history of the electric car. Energy
4. Høyer KG (2008) The history of alternative fuels in transportation: the case of electric and hybrid cars. Util Policy 16(2):63–71
5. Kumar P, Dash K (2013) Potential need for electric vehicles, charging station infrastructure and its challenges for the indian market. Adv Electr Eng 3(4):471–476

6. "Automobile industry in India." [Online]. Available: http://www.ibef.org/industry/india-automobiles.aspx
7. Ministry of Environment and Forest (2009) Results of five climate modelling studies
8. Nesamani KS (2009) Estimation of automobile emissions and control strategies in India. University of California
9. Singh SK (2006) Future mobility in India: Implications for energy demand and CO_2 emission. Transp Policy 13:398–412
10. Department of Heavy Industry (2013) National electric mobility mission plan 2020. In: Ministry of Heavy Industries & Public Enterprises, Government of India, pp 0–186
11. Mobility ET (2015) Report electric mobility FAME-India scheme—putting E-mobility on road. Springer India-New Delhi Auto Tech Review, pp 22–27
12. "Gazette Notification of FAME India." [Online]. Available: http://dhi.nic.in/writereaddata/UploadFile/Gazette_Notification_FAME_India.pdf
13. Egbue O, Long S (2012) Barriers to widespread adoption of electric vehicles: an analysis of consumer attitudes and perceptions. Energy Policy 48(2012):717–729
14. Diamond D (2009) The impact of government incentives for hybrid-electric vehicles: evidence from US states. Energy Policy 37:972–983
15. Oliver JD, Rosen DE (2010) Applying the environmental propensity framework: a segmented approach to hybrid electric vehicle marketing strategies. J Mark Theory Pract 18(4):377–393
16. Moore GA (1999) Crossing the Chasm: marketing and selling high-tech products to mainstream customers
17. Sovacool BK, Hirsh RF (2009) Beyond batteries: an examination of the benefits and barriers to plug-in hybrid electric vehicles (PHEVs) and a vehicle-to-grid (V2G) transition. Energy Policy 37(3):1095–1103
18. Axsen J, Kurani KS, Burke A (2010) Are batteries ready for plug-in hybrid buyers? Transp Policy 17(3):173–182
19. Ozaki R, Sevastyanova K (2011) Going hybrid: an analysis of consumer purchase motivations. Energy Policy 39(5):2217–2227

Thermal Management Solutions of Lithium-Ion Energy Storage Batteries for xEV Deployment in North India

Mohd Yaqzan, Yasser Rafat, Sheikh Abdullah and Mohammad Saad Alam

Abstract The Government is pushing the automotive industry in India to make a move towards hybrid and electric vehicles in future. Li-ion batteries have a higher energy density compared to lead acid batteries, which makes them better for portable applications. Furthermore, they offer 40–50% battery weight reduction and 20–30% volume reduction. The major obstacle in their implementation is large heat generation during operation. Hence, there is a need for an efficacious thermal management system. The air-cooled battery thermal management technique is in its preliminary stage. The area of liquid cooling is untouched, at least for EVs under Indian conditions. The battery thermal management technology with application in xEVs is evolving worldwide, but not much research is done pertaining to varying Indian environmental conditions. The study presented here aims at reviewing the previous works and points out the difficulties in implementing the present solutions. The first step is the problem identification where several environmental and loading conditions will be identified keeping in mind the Indian user. Next, will be the characterization of these conditions on the basis of load, conditions of road and environment. Last step will be process optimization by providing design and cooling solutions for thermal management taking into account the Indian scenario. Through this study, we want to identify the problem, characterize the situation and then attempt to provide a viable techno-commercial solution tailored for Indian scenario.

Keywords Lithium-ion · Thermal management · Electric vehicles Nanoparticles · Indian scenario · BMS · Liquid cooling

---

M. Yaqzan (✉) · Y. Rafat · S. Abdullah · M. S. Alam
CARET, Aligarh Muslim University, Aligarh, India
e-mail: yaqzanmohd@gmail.com

Y. Rafat
e-mail: yasser.rafat@zhcet.ac.in

S. Abdullah
e-mail: abdul.miet@gmail.com

M. S. Alam
e-mail: hybridvehicle@gmail.com

1 Introduction

With depleting oil reserves, electric vehicles and hybrid electric vehicles (EVs and HEVs) may present the best near-term solution for the transportation sector to reduce our dependence on petroleum and to reduce emissions of greenhouse gases and criteria pollutants. Lithium-based batteries have shown higher specific energy and specific power, thus considered a major source of electrical energy storage for the powertrains used for HEV/EVs [1]. Li-ion batteries (LIBs) have shown extended efficiency, high energy density and higher cycle life than any other batteries available in the market today [2]. For example, practical nickel–metal hydride (NiMH) batteries, which have dominated the HEV market, have a nominal specific energy and energy density of 75 Wh/kg and 240 Wh/L, respectively. In contrast, LIBs can achieve 150 Wh/kg and 400 Wh/L, i.e. nearly twice the specific energy and energy density [3].

The main barriers to the deployment of large fleets of vehicles on public roads, equipped with LIBs continue to be safety, cost (related to cycle and calendar life) and overheating during operation. These issues can strongly affect the battery performances and lifespan. In addition, more research needs to be done for transport applications of LIBs where high electric power is used in a relatively short time period. Temperature affects the longevity, efficiency and safety of the battery. Thermal runaway, electrolyte fire and, in certain cases, explosions [4] can occur when the temperature in the battery is too high. Furthermore, most of the research on these types of batteries has been related to finding the best material for the electrodes in terms of specific energy, power and cycle life, but with relatively little attention paid to thermal management [5]. This presents a significant gap in the knowledge commercial manufacturers and developers' need, to design and fabricate safe and reliable battery systems for EVs and HEVs.

The study presented here discusses previous studies made by researchers on various methods of cooling LIBs to understand the limitations of their solutions pertaining to the Indian scenario. Further, the characterization of existing problems that limit the application of xEVs locale to Indian conditions will be made in order to adequately define the thermal management problem. Thereafter, a solution will be proposed by extrapolating the prior results to lay a path for the future research.

2 Critical Review, Reported Solutions and Their Limitations

The resistance of the current collection system for the 18,650 cells is only a small fraction of the total cell impedance [6]. When this resistance is subtracted from the total cell impedance, the impedance so obtained for the bare cell layers is believed to be a reliable basis for designing full-scale cells and batteries. In designing the full-scale batteries, the calculated resistance of the cell current collection system is added to the impedance of the basic cells to obtain the total impedance. The calculated impedance for the batteries is extended to include the lumped parameter model data and is believed to be sufficiently reliable for use in vehicle simulation studies.

For EV batteries, only the battery surface temperature can be measured in real time. In a study by Zhang et al., an online battery internal temperature estimation method was proposed based on a novel simplified thermoelectric model. The experimental results confirmed the efficacy of the proposed method, and it can be used for online internal temperature estimation which is a key indicator for better real-time battery thermal management. The proposed estimation method is based only on the online measurable signals, e.g. battery voltage, current and shell temperature, and thus can be implemented in real time. Test data were collected using a LiFePO$_4$/C battery. The modelling and internal temperature estimation results confirmed the effectiveness of the proposed method [7].

2.1 Demerits of Air Cooling

Ramadass reported significant increase in capacity fading for commercial 18,650 cells when operating temperature increased from 45–50 °C. It was also clear that cell maximum temperature difference (CMTD) increased significantly with increasing flow rate, which makes the use of air cooling under high-rate discharge undesirable and may shorten battery cycle life. Adequate active cooling at high discharge rates and high ambient temperatures requires air flow rates close to or within the turbulent range which is not practical for vehicular applications [8]. While the expense of parasitic fan power under such conditions remains relatively small, the effect on non-uniformity of the cell temperatures within the pack is more appreciable and likely to impact battery cycle life. Active thermal management using a fan and air is simpler than liquid cooling systems, but may not be as effective. Since LIBs can deliver more power than NiMH batteries, they generate more heat for the same volume and thus need a more efficient heat removal system [9]. Kim and Paseran [10] compared air and liquid cooling thermal management techniques. They concluded that liquid cooling provides much better heat transfer rate.

The experimental and simulation results by Wu et al. [11] also showed that cooling by natural convection is not an effective means for removing heat from the battery system. It was found that forced convection cooling can mitigate temperature rise in the battery. Nevertheless, a non-uniform distribution of temperature on the surface of the battery is inevitable and this makes thermal management difficult. These findings suggest that forced convection cooling is capable of dissipating heat and is the reason why liquid cooling is preferred in many heat dissipation designs.

2.2 Unsuitability of PCM Cooling

Unlike most commercial batteries (e.g. lead acid, nickel cadmium and nickel–metal hydride), most commercially available Li-ion batteries exhibit a net cooling effect during charge and are highly exothermic during discharge. In a review by Hallaj et al. [12], it was shown that the technology base for a PCM incorporating battery module, operating as a passive thermal management system, is available and that the design is relatively simple, thereby promising appreciable cost reduction compared to active cooling systems. But, the drawback being PCM-based thermal management system was especially effective for batteries only under very cold ambient conditions and in space applications. During discharging, major heat flow is towards the PCM, but later, the bulk of the heat stored in the PCM is evolved towards the cell(s) as their temperature drops during charging of the battery [12] which evidently is unsuitable for Indian vehicles where the batteries are still at higher average temperatures to accept heat from the PCM.

Increasing the volume of the battery pack and amount of PCM, about doubling it [13], improved the performance of the thermal management system. But, contrary to Indian environmental conditions, the battery was operated in an environment at 30 °C while the temperature of battery pack exceeded 55 °C in some cases. Consequently, if we suppose the ambient temperature close to 40 °C, then the final resulting temperature will be much higher than the thermal runaway capacity of the battery. They also concluded that the lack of suitably sized batteries is a major obstacle in the development of successful electric vehicles [12].

3 Understanding the Indian Scenario

Before devising any system, the foremost step is to define the problem completely. Although the BTMS is a prerequisite for any application involving LIBs, the necessary operating conditions locale to Indian environment have not been characterized before.

There have been many experimental investigations that measure heat produced by lithium-ion batteries, but the majority has been on small coin cells discharged/charged at low to moderate rates at near-ambient temperatures. Very few investigators have directly measured heat rates for LIBs at rates above C/1, which routinely occur in HEV applications. Furthermore, of the studies conducted on geometrically and chemically similar cells, the results have been inconsistent. In some cases, for discharge beginning at the fully charged condition, the heat rate begins at 0 W/L, while others show a step-change to a nonzero value. Some studies also showed an approximately linear increase in heat generation as discharge proceeds, while others show it roughly constant until the end of discharge was reached [14]. The same cells tested using different methods also show significantly different results. These issues arise from difficulty in distinguishing the heat generation rate from heat storage inside the cell and have led to poor prediction of the temperature and volumetric heat generation rate at high currents. This may also be a consequence of significant variation of the heat generation rate due to different battery designs. Thermal simulation studies have provided some important insights into the operation of LIBs. As the temperature of the battery increases, the electrochemical efficiency of the battery improves, thus lowering its heat generation. In contrast, very few experimental studies have collected temperature-dependent data and those that have are restricted to narrow temperature and current ranges. The temperature dependence of heat generation causes the performance of the battery to be strongly dependent on thermal management [14]. The essential thermal problem for batteries is the poor thermal conductivity that creates a large thermal resistance between the heat generation locations and the cooling fluid. Liquid cooling and PCM cooling can improve the cell-to-cell temperature uniformity. However, the PCM increases the thermal resistance between the battery and the cooling fluid, which causes the pack temperature to rise significantly. Furthermore, although reported in the literature, the composite thermal conductivity of batteries is not well known, especially perpendicular to the stack direction. This is a critical parameter for designing an effective thermal management system. Furthermore, due to its low density and specific heat, direct cooling with air is not the best method for evenly cooling large battery packs. The increased surface area from these solutions could mitigate the temperature distribution and rise inside the battery, which should allow for more uniform discharging and charging and less thermal accumulation from cycling. This reduces the thermal resistance between the heat generation locations and the cooling media.

Some areas of research in lithium-battery thermal issues need further exploration. First, it is evident that a thermal management technique that incorporates all the features does not exist. New techniques should be explored that allow for cold storage to limit capacity fade, fast heating to improve performance, even cooling to reduce complexity in electrical balancing schemes and increasing the amount of energy extracted, and fast cooling during critical heating events.

Contrary to the initial conditions at which most of the aforementioned research work has been done, Indian conditions of temperature and climate pose a strenuous load over the xEV batteries. The major drawback is that the previous works were not sufficient to address the problems of BTM that prevail in India. There is a significant difference between the ambient temperatures of 40–45 °C here, as opposed to 20–25 °C under which most of the existing electric and hybrid electric vehicles are tested. While higher temperatures can be tolerated temporarily, if the temperature is greater than 60 °C for a prolonged period with the battery fully charged, there is a risk that the batteries may rupture, explode or catch fire due to thermal runaway. Moreover, the road conditions for which most of the present electric vehicles are being tested and are available are completely different than Indian conditions. The difference in kind of roads and the driving pattern of people in India is significantly different from rest of the world. This difference in driving pattern introduces variance in loading conditions on the LIB pack due to sudden acceleration and braking as well as overloading of the vehicle. In addition, due to untimely availability of electric supply, the BTMS will have to be designed so as to improve the performance of Li-ion cells at lower SOC conditions. The customary traffic jams and unmaintained roads produce erratic driving patterns whose effect on the battery modules in the form of vibrations, jerks, sudden temperature rise or high discharge rates has to be taken care of. For example, vehicular technology such as regenerative braking can pose a negative effect on the battery module by overcharging the battery due to habitual braking as a result of the shoddy Indian roads. Once proper characterization of these problems has been done, then only a becoming TMS that mitigates the specific battery thermal problems by addressing the local conditions can be implemented.

Finally, it can be said that the life improvement of the battery can be obtained with a specific type of TMS. And the decision to upgrade a thermal management system depends on the region where the vehicle is being used.

4 Proposed Solution

The soaring temperatures in summers in India obviate the use of active cooling and demands liquid-based cooling to ensure extended battery life and pose minimal risk to the user. The major advantages of liquid-cooled battery management are larger heat removal rate, less dependency on atmospheric temperature as the liquid can be cooled in a heat exchanger as the system is a closed one, unlike air-cooled systems. The thermal conductivity of fluids can be varied according to requirements as opposed to the use of PCM or air where the alternatives are limited. Also, direct liquid cooling and indirect liquid cooling add approximately 2.95 and 7.16% weight to the battery, respectively, which is acceptable in electric drive vehicle (EDV) applications [15].

At stressful condition, i.e. at high discharge rates and at high operating or ambient temperatures (e.g. 40–45 °C), air cooling is not a proper thermal management system to keep the temperature of the cell in the desirable operating range. The disadvantages include limiting flow rate of cooling air to 100–250 m^3/h, dependence on vehicle cabin air temperature, low cooling performance, noise, inhomogeneous temperature distribution within batteries, risk of fouling and potential safety concerns due to emission of toxic gaseous from the battery pack [9, 16–21]. The weak point that has limited widespread use of PCM system is its insufficient long-term thermal stability and low conductivity. Due to the low conductivity, the PCM cooling system will suffer from the inherent limitations such as the moderate temperature gradient and long cooling time. Although many matrices were developed to improve the thermal conductivity of the PCM, the high cost and volume change of PCM cannot be well managed [22]. Liquid cooling works well because it allows relocation of heat from source to a more efficient location for cooling. The liquid acts more of a heat carrier, and liquid cooling is designed to remove heat from areas where air is hard to reach and ventilation is hard to introduce.

Moreover, properties such as thermal conductivity and heat capacity of coolants can be enhanced to serve the purpose of BTM by addition of suitable substances to water or water-based liquids, while simultaneously, ensuring to keep the viscosity and density of liquid to minimum to avoid extra pumping power. The prismatic Li-ion cells will be more suited for EV application as the prismatic cell's thin, rectangular shape facilitates better layering and gives product designers increased flexibility. They have large surface area-to-volume ratio as compared to typical cylindrical cells which facilitates greater heat dissipation from the cell's surface.

5 Simulation and Experimentation

5.1 2D Simulation in a Prismatic Cell Using the Air as Cooling Medium at Different Heat Flux (Initial Assumption) and Reynolds Number

Dimension of fluid boundary—2 mm × 31 mm
Input parameters—flow from single side wall (Figs. 1, 2 and 3)

At heat flux of 500 w/m^2 and inlet air velocity changing from 2 to 15 m/s, the maximum temperature decreases but it converges instantly as compared to results given by other scientists. So we analyse that the heat flux is much greater than used in these simulations. All these results are found at normal atmospheric condition. But as is the case of atmospheric temperature close to 45 °C during Indian summers, air cooling is not able to control wall temperature.

Fig. 1 Heat flux—500 w/m^2, inlet velocity—2 m/s

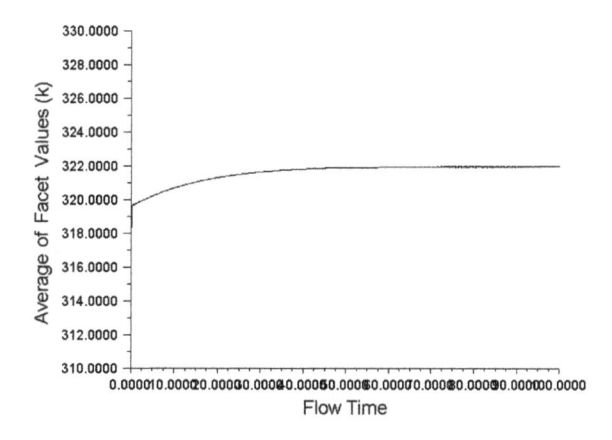

Fig. 2 Heat flux—500 w/m², inlet velocity—10 m/s

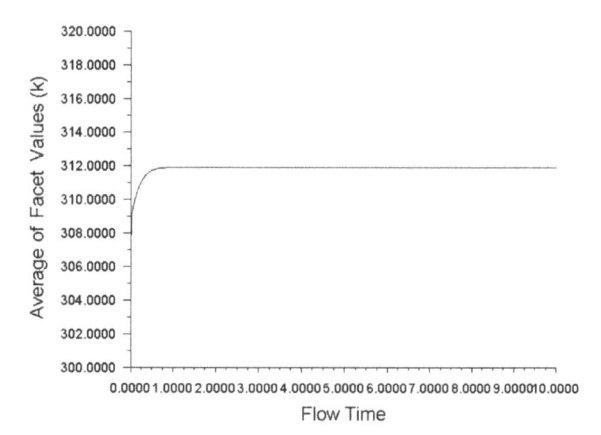

Fig. 3 Heat flux—500 w/m², inlet velocity—15 m/s

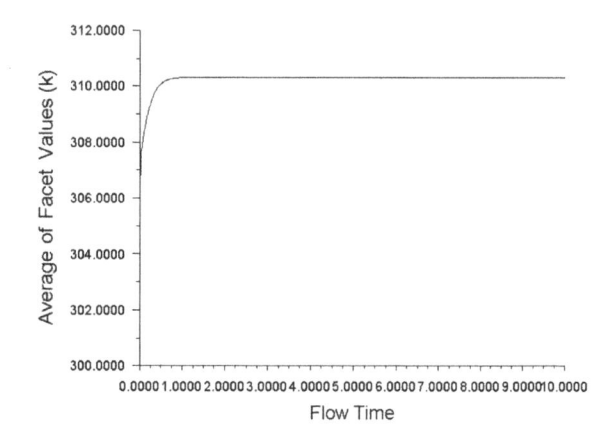

5.2 2D Simulation in a Prismatic Cell Using Water as Cooling Medium

Dimension of fluid boundary—3 mm × 131 mm
Input parameters—heated from both side (Figs. 4, 5, 6, 7 and 8)

5.3 2D Simulation in a Prismatic Cell Having Constriction Shape Using Water as Cooling Medium

The above simulations were performed in lieu of the experimental set-up being made for testing and controlling the lithium-battery heat generation. Moreover, synthesis of a nanofluid is also performed to meet the requirements of rise in

Thermal Management Solutions of Lithium-Ion … 187

Fig. 4 Heat flux—10,000 w/m^2, velocity—0.02 m/s

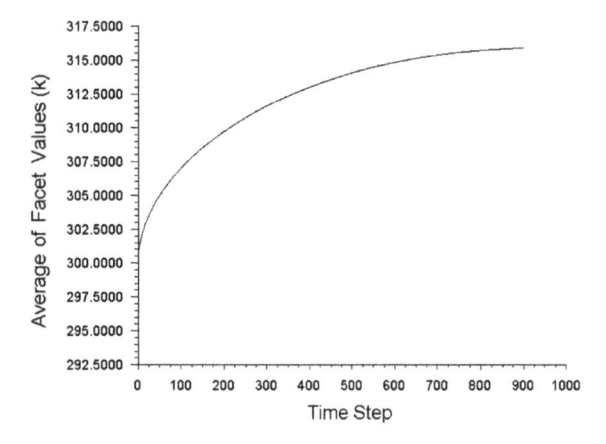

Fig. 5 Heat flux—10,000 w/m^2, velocity—0.05 m/s

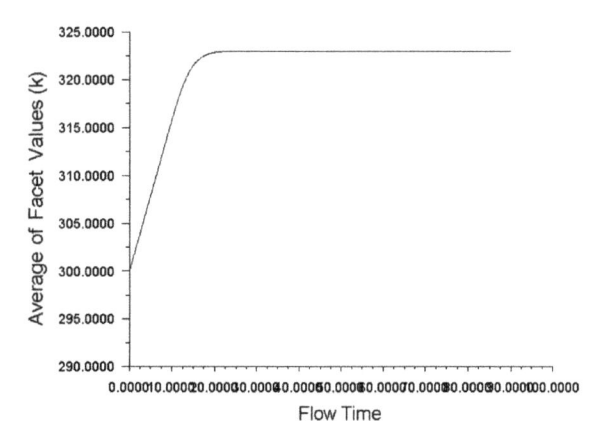

Fig. 6 Heat flux—10,000 w/m^2, velocity—0.06 m/s

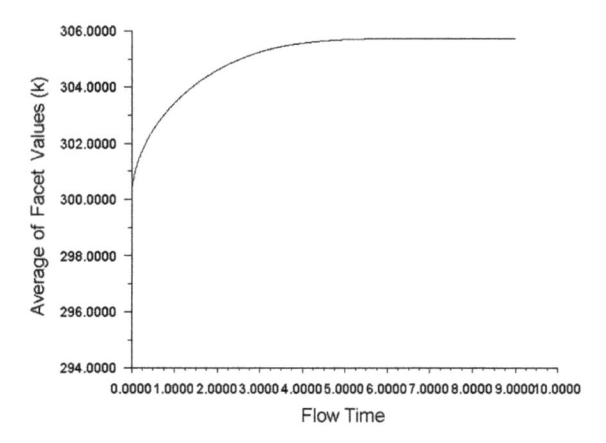

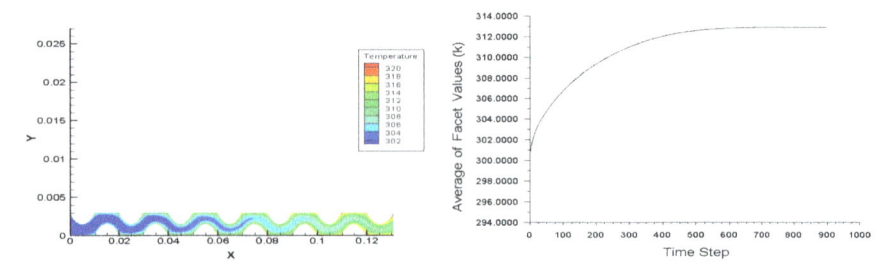

Fig. 7 Heat flux—10,000 w/m², velocity—0.02 m/s

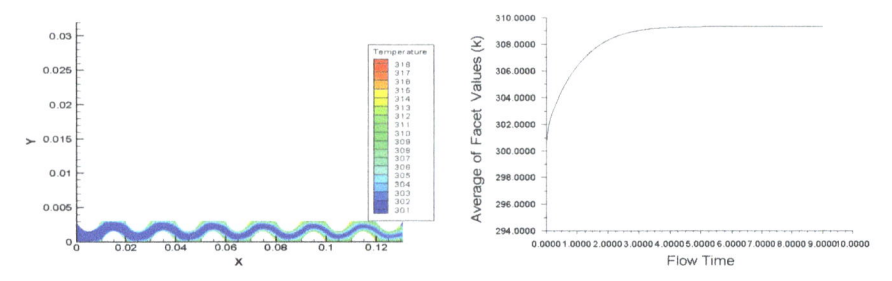

Fig. 8 Heat flux—10,000 w/m², velocity—0.05 m/s

Fig. 9 SEM image of TiO$_2$ nanoparticle at 15,000× magnification

thermal conductivity beyond exception, ultrafast heat transfer ability, better stability than other colloids, reduction of clogging in microchannels, reduction in pumping power and reduced friction coefficient (Fig. 9).

Jang and Choi [23] model is one of the latest developments based on Brownian motion. The mathematical model is expressed as follows:

$$K_{\text{eff}} = K_{\text{f}} + (1 - \phi) + K_{\text{nano}}\phi + \phi\, h\, \delta_{\text{T}},$$

where K_{nano} is the thermal conductivity of nanoparticle, h is coefficient of heat transfer, δ_{T} is the thickness of layer and ϕ is particle volume fraction. Ethylene glycol–water mixture and engine oil are very poor heat transfer fluids. Even they perform worse than water. Replacement of such poor cooling medium with nanofluids can be fruitful for automotive cooling system. Such improvement can be used to remove heat from relatively smaller size of coolant system. Smaller coolant system results in smaller and lighter radiators which lead to decrease engine weight to some extent. Lighter engine component can also increase fuel efficiency in cars and trucks [24]. This nanofluid along with the developed experimental set-up will help to provide a customized solution for electric vehicles for Indian roads, and its results will be discussed in the future work.

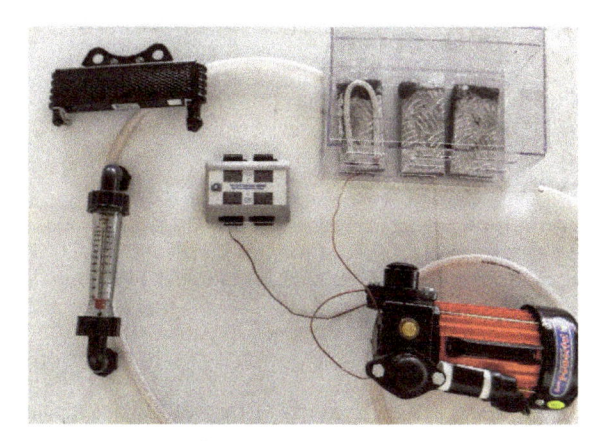

Hardware setup model for BTMS

6 Conclusions

The advantages and disadvantages of various cooling methodologies were discussed based on the previous researches. It was seen that air cooling is not a proper thermal management system to keep the temperature of the cell in the desirable operating range at high operating or ambient temperatures. Also, the PCM cooling system will suffer from limitations such as insufficient long-term thermal stability and low conductivity. Very few experimental studies have collected temperature-dependent data and those that have are restricted to narrow temperature and current ranges.

The conditions prevailing in India were presented that make the implementation of Li-ion batteries a completely different and difficult situation than those studied so far. A solution to these problems was also proposed in the form of liquid cooling, and various parameters were discussed that enhance the cooling capacity of BTMS in future. Also, once adequate amount of experimental data will be available speaking as of Indian environment, it will be easy to simulate those conditions in the laboratory. Even then, several simulation results pertaining to different geometries were performed to compare the effects of varying pack geometry on cooling behaviour. These data together with the use of numerical solutions will also provide simulation results that would be useful to battery pack designers of electrical vehicle to assess and choose a proper cooling method under the volumetric constraints.

Although simulation results gave promising data for future research, hardware results will give a better insight into the actual problems and temperature abnormalities observed pertaining to Indian scenario. Hence, the hardware set-up so far is in accordance with the data obtained from previous literature research.

References

1. Doucette RT, McCulloch MD (2011) A comparison of high-speed flywheels, batteries, and ultracapacitors on the bases of cost and fuel economy as the energy storage system in a fuel cell based hybrid electric vehicle. J Power Sour 196(3):1163–1170
2. Gerssen-Gondelach SJ, Faaij AP (2012) Performance of batteries for electric vehicles on short and longer term. J Power Sour 212:111–129
3. Linden D, Reddy TB (2002) Handbook of batteries, 3rd. McGraw-Hill
4. Lu L et al (2013) A review on the key issues for lithium-ion battery management in electric vehicles. J Power Sour 226:272–288
5. Alavi-Soltani S, Ravigururajan T, Rezac M (2006) Thermal Issues in Lithium-Ion Batteries. In: ASME 2006 international mechanical engineering congress and exposition, 2006. American Society of Mechanical Engineers
6. Nelson P et al (2002) Design modeling of lithium-ion battery performance. J Power Sources 110(2):437–444
7. Zhang C, Li K, Deng J (2016) Real-time estimation of battery internal temperature based on a simplified thermoelectric model. J Power Sour 302:146–154
8. Pesaran AA (2001) Battery thermal management in EV and HEVs: issues and solutions. Battery Man 43(5):34–49
9. Sabbah R et al (2008) Active (air-cooled) versus passive (phase change material) thermal management of high power lithium-ion packs: limitation of temperature rise and uniformity of temperature distribution. J Power Sour 182(2):630–638
10. Kim G-H, Pesaran A (2006) Battery thermal management system design modeling. National Renewable Energy Laboratory
11. Wu M-S et al (2002) Heat dissipation design for lithium-ion batteries. J Power Sour 109(1):160–166
12. Al-Hallaj S, Selman J (2002) Thermal modeling of secondary lithium batteries for electric vehicle/hybrid electric vehicle applications. J Power Sour 110(2):341–348
13. Mills A, Al-Hallaj S (2005) Simulation of passive thermal management system for lithium-ion battery packs. J Power Sour 141(2):307–315

14. Bandhauer TM, Garimella S, Fuller TF (2011) A critical review of thermal issues in lithium-ion batteries. J Electrochem Soc 158(3):R1–R25
15. Chen D et al (2016) Comparison of different cooling methods for lithium ion battery cells. Appl Therm Eng 94:846–854
16. Van den Bossche P et al (2006) SUBAT: An assessment of sustainable battery technology. J Power Sour 162(2):913–919
17. Omar N et al (2012) Standardization work for BEV and HEV applications: critical appraisal of recent traction battery documents. Energies 5(1):138–156
18. Krieger A, Rathke P, Wang L (2012) Recharging China's electric vehicle aspirations. Retrieved 10:2012
19. Saw L, Tay A, Zhang LW (2015) Thermal management of lithium-ion battery pack with liquid cooling. In: Thermal measurement, modeling and management symposium (SEMI-THERM), 31, 2015, IEEE
20. Pesaran AA (2002) Battery thermal models for hybrid vehicle simulations. J Power Sour 110(2):377–382
21. Saw L, Tay A (2013) Thermal modeling and management of Li-ion batteries for electric vehicles. In: Proceedings of the ASME 2013 international technical conference and exhibition on packaging and integration of electronic and phonic microsystems, 2013
22. Zhao R, Gu J, Liu J (2015) An experimental study of heat pipe thermal management system with wet cooling method for lithium ion batteries. J Power Sour 273:1089–1097
23. Jang SP, Choi SUS (2004) Role of Brownian motion in the enhanced thermal conductivity of nanofluids. Appl Phys Lett 84(21):4316–4318
24. Wang X-Q, Mujumdar AS (2008) A review on nanofluids—Part II: experiments and applications. Braz J Chem Eng 25(4)

Electric Vehicle Charging Infrastructure in India: Viability Analysis

Wajahat Khan, Furkan Ahmad, Aqueel Ahmad, Mohammad Saad Alam and Akshay Ahuja

Abstract Rising pollution is a cause of major concern for the Indian cities. Extensive reliance on IC engine-based vehicles as the principal means of transport has raised serious environmental concerns. To address the ongoing problems, there is increased emphasis on developing high-efficient, emission-free means of transport. Keeping this in view, electric and hybrid vehicles appear to be the best alternatives for replacing the conventional vehicles. But wide adoption of electric vehicles will require residential as well as public charging infrastructure analogues to petrol pumps. Level 1, Level 2, and fast EV charging stations have been installed in various countries in the world. A detailed exploration is required to develop a similar customized infrastructure for the Indian market. This work provides a detailed overview of the various technologies and viable options in terms of power quality, power electronic converter topologies, and energy management options optimally feasible in India. Recent policies and initiatives taken by the Government are also presented in this paper.

Keywords Charging infrastructure · Electric vehicles · Smart grid Smart charging management

---

W. Khan (✉) · A. Ahmad · M. S. Alam
Centre of Advanced Research in Electrified Transportation (CARET),
AMU, Aligarh, India
e-mail: wajahat.kkhan786@gmail.com

A. Ahmad
e-mail: aqueel.100@gmail.com

M. S. Alam
e-mail: saad.alam@zhcet.ac.in

F. Ahmad
Department of Electrical Engineering, Aligarh Muslim University,
202002 Aligarh, India
e-mail: furkanahmad@zhcet.ac.in

A. Ahuja
ISGF, New Delhi, India
e-mail: akshay@indiasmartgrid.org

© Springer Nature Singapore Pte Ltd. 2018
R. K. Pillai et al. (eds.), *ISGW 2017: Compendium of Technical Papers*, Lecture Notes in Electrical Engineering 487, https://doi.org/10.1007/978-981-10-8249-8_17

1 Introduction

The ongoing smart city endeavor in India has to address challenges including air pollution, increasing greenhouse gas emissions, and rising risks of energy security. Many cities in India have extremely high levels of urban air pollution in the form of oxides of carbon, sulfur, and nitrogen [1].

Transport sector is the major contributor to the environmental pollution accounting for about 51% of pollution of India as shown in Fig. 1, and in urban areas, this figure goes to 75–80%. There is growing interest in alternative fuel vehicles, particularly in electric vehicles (EVs) to minimize the negative impact of transport sector on environment. EVs have not gained wide acceptance among customers in the past. However, technological advancements, particularly in the field of battery technology, have made EVs attractive to the customers. EVs have started penetrating the automotive market in many Indian cities. However, large-scale deployment of EVs in India largely depends on the availability of EV charging infrastructure which includes slow charging stations (Level 1/Level 2) for home charging and fast charging stations situated in public places for commercial charging.

2 Building Blocks of EV Charging Infrastructure

EV charger is an integral part of the charging infrastructure. Charging time of EV is closely associated with the charger characteristics. Charger should be efficient, with high power density, low weight, and low volume. An EV charger must draw utility current with low distortions so that there is minimum impact on the power quality of grid. With conventional rectifiers, a large proportion of low-order harmonics are injected in the source current due to which power factor decreases. Poor power factor leads to inefficient utilization of volt-ampere rating. To mitigate these adverse effects, power factor correction circuits are used with the converters. Numerous

Fig. 1 Air pollution caused by different sectors in India [2]

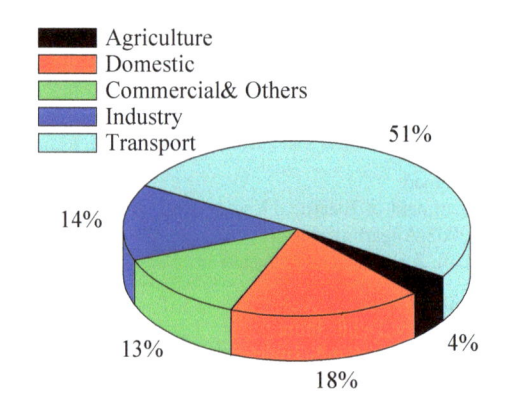

Table 1 EV converter topologies [4]

Topology	Conventional boost	Phase-shifted semi-bridgeless	Interleaved	Bridgeless interleaved
Power capacity	<1 kW	<3.5 kW	<3.5 kW	>5 kW
Output ripples	High	Medium	Low	Low
Input ripples	High	Medium	Low	Low
Efficiency	Poor	Best	Fair	Best
Cost	Low	Medium	Medium	High

control techniques and circuit topologies have been developed to achieve power factor correction. Table 1 gives the comparison between different converter topologies on the basis of power rating, capacitor ripples, input distortion, efficiency, and cost.

2.1 Power Levels of Charging

On the basis of charging power levels, EV charging can be classified as Level 1 charging, Level 2 charging, Level 3 charging, and DC fast charging.

Level 1 charging is the slowest mode of charging which requires 120 V/15 A [3], single-phase supply and J1772 EV connector to connect with the EV port. It usually takes place at home or garage. Level 2 charging is used for both public and private applications. It uses dedicated supply equipment which is connected to 208 or 240 V outlet. Level 3 and fast DC charging are used for commercial applications which take place at public places like highways, city fueling points, similar to gas stations.

Table 2 EV charging levels

Charging type	Level 1	Level 2	DC fast
Charging time (h)	20–22	6–8	0.2–0.5
Charger location	Onboard (1-phase)	Onboard (1 or 3-phase)	Off-board (3-phase)
Voltage supply (V)	120	240	208–600
Power level (kW)	1.3–1.9	Up to 19.2	50–150
Range	2–5 miles per hour of charging	10–20 miles per hour of charging	60–80 miles in <30 min
Primary use	Residential charging	Residential and public charging	Public charging

Table 2 gives the description of various charging levels based on charging time, power and voltage levels, range obtained in per hour of charging, and charger location.

2.2 Standards of EV Charging

Organizations such as International Electrotechnical Commission (IEC), Society of Automotive Engineers (SAE), Institute of Electrical and Electronics Engineers Standards Association (IEEE-SA), Japan Electric Vehicle Standards (JEVS), Infrastructure Working Council (IWC) have prepared codes and standards for development of EV charging infrastructure. Some of the standards are listed in Table 3.

2.3 Onboard and Off-board EV Charging Topologies

EV chargers can be classified as onboard and off-board chargers, with unidirectional and bidirectional power flow capabilities. A charger with bidirectional capability allows charging from the grid as well as energy injection back to the grid.

Onboard chargers are restricted by size and weight limitation due to limited space on the vehicle, while there is no such limitation for off-board chargers.

Table 3 EV charging standards [6]

Standard	Specification
SAE-J1772	EV coupler for conductive charging
SAE-J1773	EV inductively coupled charging
SAE-J1797	Recommended practice for EV battery modules packaging
SAE-J2288	Life cycle testing of battery modules for EV
SAE-J2464	EV/HEV rechargeable energy storage system (RESS) safety and abuse testing
SAE-J2836 Part 1	Use cases for communications between PEVs and utility grid
SAE-J2836 Part 2	Use cases for communications between PEVs and supply equipment (EVSE)
SAE-J2836 Part 3	Use cases for communications between plug-in vehicles and the utility grid for reverse flow
SAE-J2894	Power quality requirements for plug-in vehicle chargers—requirements
IEC-69/156/CD:2008	Electric vehicle conductive charging system
IEC-23H/222/CD:2010	Plugs, socket outlets, vehicle couplers, and vehicle inlets—conductive charging of EVs
JEVS-C601:2000	Plugs and receptacles for EV charging
AIS-138 (Draft)	Electric vehicle conductive AC charging system-ARAI

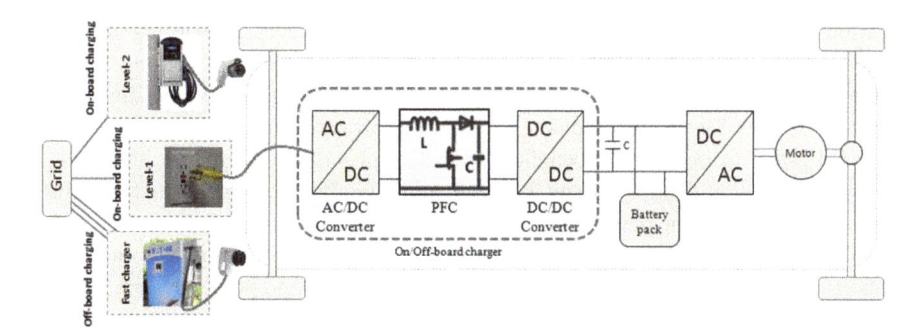

Fig. 2 Typical layout of EV charging system with on/off-board charging strategies

Off-board chargers are used for fast charging, while Level 1 charging and Level 2 charging schemes extensively use onboard chargers [5].

Figure 2 shows a typical layout of EV charger with onboard and off-board charging schemes for different levels of charging. In case of fast charging, the charger assembly consisting of AC/DC and DC/DC converters is shifted off-board and rectified output of the converter is given to the EV charging inlet.

3 Recommended Planning of EV Charging Infrastructure

Market penetration of EVs is strongly dependent on the charging infrastructure availability. Hence, development of infrastructure for EV charging is considered as the foremost task by concerned governments and electric vehicle developers. An important factor which is to be considered in this regard is that the existing power system must be capable of accommodating this new charging load. For that, an approach for charging that is consumer convenient and at the same time does not degrade the power grid is required.

Availability of commercialized charging stations would offer certain benefits to the power grid. Some of them are:

- They will help in diverting the peak of charging load from the demand peak of the network.
- Unpredictable mobile load in the form of EVs would be transformed into a stationary load, and it would be easier to predict.
- When in the form of bulk charging load, it would be simpler to enforce regulations on harmonics and power factor.
- Implementation of V2G concept would be easy as it would eliminate the need for integration of sophisticated devices for measurement, communication, and control, up to end consumer level.

Moreover, when compared with slow Level 1 and Level 2 charging modes, which take considerable time, fast charging at public charging stations would definitely be a preferable choice for EV consumers too. Hence if prior actions are taken by utilities for the development of charging infrastructure well before in time, it would prove to be of great benefit. For that, appropriate planning is important, incorporating identification of suitable sites and charging load size, satisfying the required constraints for planning.

A more realistic approach for planning would be to select locations for charging stations which satisfy the requirements of both consumers and power system. From that perspective, the location which satisfies the given criteria is considered to be ideal.

- The location should be in the vicinity of EV consumers.
- The location should be so chosen that it covers the productive EV sites like town centers, office complexes, residential areas, road networks.
- Location of charging station should be such that the regulatory limits on voltage and on line MVA flows are not violated and least losses are incurred in the power system.

Numbers of optimization techniques like genetic algorithm, simulated annealing, virtual ant algorithms, particle swarm optimization are used for the planning of charging infrastructure. Particle swarm optimization is used in [7] using an IEEE 16 bus distribution test system.

4 Cost Estimation of Charging Infrastructure

Total investment cost required for the establishment of charging infrastructure for EVs includes the cost of equipments to be used, installation costs, and operation and maintenance costs. With increase in penetration of EVs in the next few years, number of EV chargers will increase and hence the equipment cost is expected to decrease. There is large variation in the EV charging equipment cost among different manufacturers. Besides the initial equipment cost, installation cost is required for installation of the charger and interconnection with grid. The installation cost includes cost of civil works, transaction cost regarding distribution system operator permission, and other related costs depending on factors like requirement of a new grid connection or upgradation of the existing connection. In case there is a preexisting connection, the installation cost may greatly reduce. Some other parameters that influence installation cost include number of simultaneous installation of many chargers which may reduce cost on account of common labor and grounding costs, mutual components, etc.

For semiprivate/semipublic places where low- or medium-power level chargers are required, cost varies between 500 € (Rs. 36,431) and 1200 € (Rs. 87,435) [8, 9]. For public places where high power level chargers are required, installation cost is relatively higher ranging between 2400 € (Rs. 1,74,871) to 3600 € (Rs. 2,62,306) [9]. Apart from the equipment and installation costs, there is a requirement for

Fig. 3 Estimated installation cost of EV charging stations

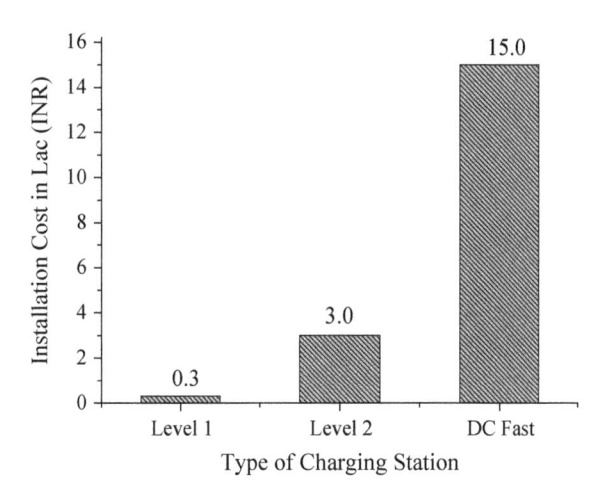

continuous maintenance over the running period. For that, operation and maintenance cost is to be added, which may be taken to be 10% of the total installation cost (including equipment cost) [10].

Figure 3 shows the cost of installing EV charging stations in Delhi as estimated in [11] for Level 1 charger of 1.5 kW, Level 2 charger of 6.6 kW, and DC fast charger of 50 kW capacity.

5 Impact on Power Grid

There is large randomness in the behavior of EV charging, which is greatly influenced by factors like mileage, battery capacity, charging time. Interconnection of charging infrastructure with the grid has some effects on the distribution network which are discussed here.

- The first issue that comes on interconnecting charging infrastructure is that it increases the difficulty of distribution network planning. New constraints in the form of electricity demand and the layout of charging stations need to be considered, which add to the complexity of network planning.
- Improved quality equipments with high ratings are required in the distribution network to facilitate interconnection of charging infrastructure. Integration of high-power charging station requires distribution transformer with larger capacity and distribution line of larger cross section to avoid problems like overloading, voltage deviation.
- It may lead to a decrease in the economy of distribution system operation. As charging load exhibits large volatility, it is difficult to confine charging behavior to low-load periods, leading to greater system peak difference. This would ultimately result in lower utilization efficiency of distribution network equipment.

- Power quality of the distribution network is affected. Charger uses several power electronic conversion devices in the form of converters which induce harmonics in the source-side current.

6 Need for Smart Charging Management

With increasing EV penetration in the Indian scenario, the demand for charging is expected to increase proportionally, and it would act as an additional burden on the grid. This additional load is bound to affect the grid adversely, if optimal scheduling is not done in advance. It will lead to imbalance of voltage as well as frequency, which ultimately may lead to grid failure and blackout. Continuous monitoring of power system is necessary while charging large number of EVs, in order to achieve grid balancing. The concept of smart grid, wherein a two-way communication can be established between the utility and consumers, has to be reinforced. A typical demand profile for one day, obtained from Power Exchange (IEX), is shown in Fig. 4 which is a replica of the energy demand in Indian scenario, where minimum load is observed during the night hours and also during mid of the day.

Low peak hours during night, wherein the load is at its minimum, are most suitable for home-based charging, using either Level 1 or Level 2. This period of 6–8 h is also very good for charging batteries at battery swapping stations (BSS). Public charging stations (CSs), where most of the charging is expected to occur during daytime, can shift their charging load to low peak hours to enhance the functioning of grid. This can be done by deploying attractive charging schemes for the consumers during low peak hours or by storing energy during this period as a backup for peak load hours. This type of scheduling can prove to be beneficial for both the utility and consumers. A flatter load profile with less system peak differences can be obtained this way, which is desirable. Also, from the consumer's

Fig. 4 Demand profile for one day

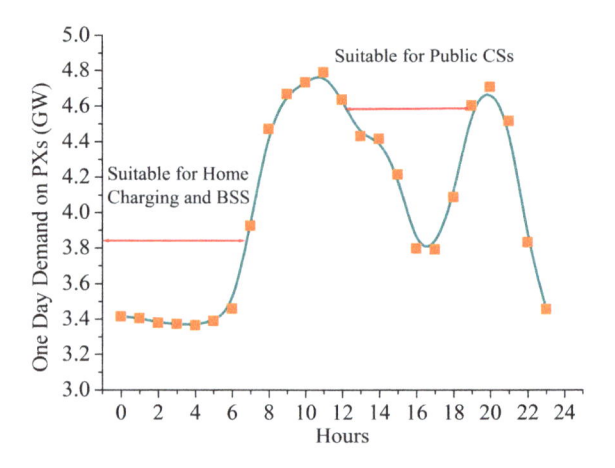

Fig. 5 Cost of per unit electricity

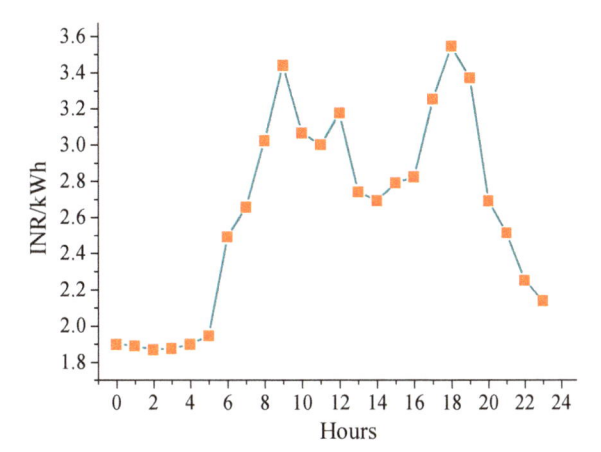

viewpoint, this is advantageous as evident from Fig. 5. Cost of purchasing electricity is lower during low peak hours as compared to that during peak hours.

Such a demand-side management can help in smooth running of power grid with fewer disturbances. Further enhancement of grid operation can be achieved by injection of power back to the grid using vehicle to grid (V2G) concept and also by participating in reserve market to maintain the frequency up and down regulations. This can prove to be beneficial for the EV aggregator as well as to consumers, in terms of economy.

The aggregator of EV charging station as well as BSS can gain maximum profit by optimizing the cost function of trading revenue mentioned in Eq. (1). An optimization model based on this cost function is given in [12].

$$\max\{r^{em} + r^{cap} - c^{regup} - c^{regdn}\} \quad (1)$$

where r^{em} represents the day-ahead (DA) revenue of energy market, r^{cap} represents the DA regulation market revenue for the capacity, c^{regup} represents regulation up service cost, and c^{regdn} represents regulation down service cost.

The DA revenue of energy market r^{em} can be expressed as:

$$r^{em} = \Delta t \sum_{t \in T} \sum_{v \in V} \lambda_t^{DA}(\eta_v^{dsg} \cdot p_{t,v}^{emdsg} - p_{t,v}^{emchg}) \quad (2)$$

where λ_t^{DA} is DA energy market price, η_v^{dsg} is battery discharge efficiency, $p_{t,v}^{emdsg}$ and $p_{t,v}^{emchg}$ represent discharging and charging powers, respectively.

The revenue r^{cap} is obtained as:

$$r^{cap} = \sum_{t \in T} \sum_{b \in B} [(w_{t,b}^{up} \lambda_{t,b}^{up}) \pi^a p_t^{up} + (w_{t,b}^{dn} \lambda_{t,b}^{dn}) \phi^a p_t^{dn}] \quad (3)$$

where $\lambda_{t,b}^{up}$ and $\lambda_{t,b}^{dn}$ are DA regulation up and down capacity prices, respectively; power p_t^{up} and p_t^{dn} are regulation up and down capacities offer to the market; $\pi^{a/d}$ is the probability of acceptance and deployment, respectively, for up regulation; $\phi^{a/d}$ is the probability of acceptance and deployment, respectively, for down regulation; $w_{t,b}^{up/dn}$ and $v_{t,b}^{up/dn}$ are the segment activation of the price quantity for capacity and real-time energy probability curves in period t and segment b.

Cost for regulation up service (c^{regup}) can be expressed as:

$$c^{regup} = \pi^a \pi^d (1 - \pi^d) \sum_{t \in T} \sum_{b \in B} (v_{t,b}^{up} \lambda_{t,b}^{RT})(p_t^{up} - \pi^a \pi^d p_t^{up}) \tag{4}$$

Similarly, cost for regulation down service is expressed as:

$$c^{regdn} = \Phi^a \Phi^d (1 - \Phi^d) \sum_{t \in T} \sum_{b \in B} (v_{t,b}^{dn} \lambda_{t,b}^{RT})(p_t^{dn} - \Phi^a \Phi^d p_t^{dn}) \tag{5}$$

In order to participate in DA market, proper load forecasting is essential. Numerous techniques have been developed for load forecasting like simplistic benchmark methods, seasonal ARMA modeling, periodic AR models [13]. For simplistic benchmark method, the forecast function is given as:

$$\overline{y_{(t)}}(k) = y_{t+k-s_2} \tag{6}$$

where y_t represents the demand in period t and k is the forecast lead time ($k \leq s_2$). Second simplistic benchmark does simple averaging of the corresponding observations in the past for forecasting. The forecast function for this method is given in Eq. (7) for four observations.

$$\overline{y_{(t)}}(k) = (y_{t+k-s_2} + y_{t+k-2s_2} + y_{t+k-3s_2} + y_{t+k-4s_2})/4 \tag{7}$$

The uncertainties involved in the EV fleet characteristics, DA electricity market operations as well as in generation, transmission, and load may be represented by Monte Carlo simulation. Risk involved with the financial as well as economical aspects of the EV aggregator in uncertainty environments can be managed by conditional value-added risk analysis (CVaR) as given in Eq. (8).

$$\text{Maximize}_{\varsigma, \delta^s} \text{ CVaR} = \varsigma - \frac{1}{(1-a)} \sum_{s=1}^{NS} \text{Pr}^s \cdot \delta^s \tag{8}$$

For detailed risk analysis based on the above method, readers may go through [14].

7 Key Barriers and Potential Solutions

Scaling up EV penetration in Indian automotive market requires addressing certain challenges and barriers related to infrastructure, market, technical limitations, and policies. Technical challenges include dealing with issues like improvement in battery technology, reduction in charging time, and getting sufficient driving range for EVs. Low specific energy density of EV batteries like lead–acid batteries is a major barrier. So to achieve higher driving ranges, bulky batteries are required, which increase the overall weight of the vehicle. To overcome this issue, battery makers are working on developing batteries having high values of specific energy density. One major concern is that India has lack of lithium-ion reserves, which give high energy density batteries [15].

Market and infrastructural barriers are related to lack of charging infrastructure, absence of business models to meet specific needs of EVs, and lack of dedicated lanes. Moreover, commercial stations for charging and battery swapping should be there at public places to allow ease of charging. Similar to technical barriers, solutions exist for market and infrastructural barriers also. To address the problem of charging infrastructure, a solution could be to set up charging points in parking garages and in the basements of buildings. This strategy has been adopted by many countries like China. There is also a need to build business models which may reduce upfront investment of consumers and increase consumer confidence in electric vehicles.

Lack of well-defined policy framework for the promotion of EVs in the country is a major barrier to widespread adoption of EVs. Policymakers need to implement a policy road map keeping in view the benefits provided by EVs. Governments need to support research activities in battery technology and vehicle innovations, along with providing incentives to enhance EVs production and sale. There is also a need to understand the public perceptions regarding safety, driving range, and cost of these vehicles compared to conventionally powered vehicles and make policies in light of them. National policies like NEMMP may prove to be a major step toward this goal.

8 EV Market in India

The present EV market in India is not so large. Though there are plenty of electric two-wheelers and four-wheelers in the market, but even then EVs' overall share is negligible. Electric vehicles started penetrating the Indian automotive market long back in the 1990s, when electric two- and three-wheelers were first launched. But they did not prove to be successful and were later discontinued. A major breakthrough came in the early 2000s with the launch of first electric car in India by Mahindra Electric, named Reva. Mahindra Electric emphasized on developing electric cars at reasonable prices. It launched its first model in India in the year 2001

[16]. A modified version of this car came in the market in the year 2013. This new model used lithium-ion powered batteries and allowed for driving range of 100 km with full charge. Prius hybrid model was launched by Toyota in Indian market in the year 2010. This was followed by Camry Hybrid in 2013. In few Indian cities, hybrid and electric buses have been introduced. Municipal Transport Corporation of Bangalore launched an electric bus, few years back. Electric rickshaws have emerged as excellent substitutes for the conventional three-wheeled vehicles as well as for the paddle rickshaws. But still there is vast scope for the expansion of EV market in the country.

9 Policy Framework in India

Several initiatives have been taken by the government–industry collaboration to promote the widespread adoption of EVs in the country. "Ministry of New and Renewable Energy (MNRE)" has promoted the R&D work on electric vehicles as part of the Alternative Fuel for Surface Transportation program. "National Electric Mobility Mission Plan (NEMMP) 2020" [17] has been launched by Department of Heavy Industry for the promotion of EVs in India. The mission aimed at deploying 6–7 million electric vehicles in India by the year 2020. A scheme has been formulated named "FAME India (Faster Adoption and Manufacturing of (Hybrid and) Electric Vehicles in India)" [18], as a part of this mission. This scheme is to be implemented over 6-year period, till year 2020. This scheme focuses on four major areas including technological development, charging infrastructure, pilot projects, and demand creation. The percentage breakup of total fund (Rs. 795 crore) allocated to key areas is given in Fig. 6.

Fig. 6 Fund allocation under FAME India scheme [19]

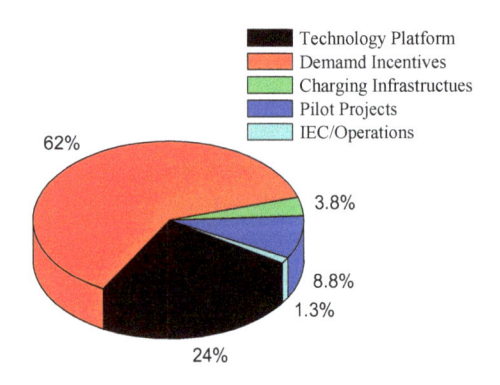

10 Conclusion

India ought to put resources into little scale fortifications to deal with the electric load issues locally, rather than going for an enormous change. Home charging ought to be keyed up for long battery life and framework adjustment. Legitimate arrangement of place, populace, traffic density behavior, and the security ought to be thought about before going for a large-scale charging infrastructure execution for the second biggest populated nation of the world. The most essential aspect in the working of an electric vehicle charging framework is the incorporation of exercises in the electrical vitality and transport fields. Just deliberate improvement of both frameworks will give steady and dependable electrical power framework, particularly at the level of restricted power and feasible advancement of the Indian electric vehicle advertise. Development of V2G concept along with G2V implementation on large scale would help in dealing with grid-related issues.

Acknowledgements This research work has been supported by the Centre of Advanced Research in Electrified Transportation (CARET), Aligarh Muslim University, Aligarh, India.

References

1. Shrivastava RK, Neeta S, Geeta G (2013) Air pollution due to road transportation in India : a review on assessment and reduction strategies. J Environ Res Dev 8(1):69
2. "India-One of the world's most exciting EV market" www.un.org/esa/dsd/susdevtopics/sdt_pdfs/meetings2012/statements/anupam.pdf
3. Yilmaz M, Krein PT (2012) Review of charging power levels and infrastructure for plug-in electric and hybrid vehicles. In: 2012 IEEE International Electric Vehicle Conference. Greenville, SC, USA, pp 1–8
4. Prabhakar S, Febin Daya JL (2016) A comparative study on the performance of interleaved converters for EV battery charging. In: 2016 IEEE 6th International Conference on Power Systems (ICPS). New Delhi, pp 1–6
5. Yilmaz M, Krein PT (2013) Review of battery charger topologies, charging power levels, and infrastructure for plug-in electric and hybrid vehicles. IEEE Trans Power Electron 28(5):2151–2169
6. Foley AM, Winning IJ, Gallachóir BPÓ (2010) State-of-the-art in electric vehicle charging infrastructure. In: IEEE Vehicle power and propulsion conference (VPPC), 2010
7. Dharmakeerthi CH, Mithulananthan N, Saha TK (2013) Planning of electric vehicle charging infrastructure. In: 2013 IEEE Power & Energy Society General Meeting. Vancouver, BC, Canada, pp 1–5
8. Vollers W, Steinmetz R, Han Q (2013) Sustainable business models for public charging points. In: 2013 World Electric Vehicle Symposium and Exhibition (EVS27). Barcelona, pp 1–11
9. Jahan S, Habiba R (2015) An analysis of smart grid communication infrastructure & cyber security in smart grid. In: 2015 International Conference on Advances in Electrical Engineering (ICAEE). Dhaka, pp 190–193
10. Vagropoulos SI, Kleidaras AP, Bakirtzis AG (2014) Financial viability of investments on electric vehicle charging stations in workplaces with parking lots under flat rate retail tariff

schemes. In: 2014 49th International Universities Power Engineering Conference (UPEC). Cluj-Napoca, pp 1–6
11. Clean enenrgy ministrial (2014) Assessing and Accelerating Electric Vehicle Deployment in India, May 2014. Available at www.cleanenergyministerial.org/Portals/2/pdfs/EVILBNL_India_May2014report.pdf
12. Sarker MR, Dvorkin Y, Ortega-Vazquez MA (2016) Optimal participation of an electric vehicle aggregator in day-ahead energy and reserve markets. IEEE Trans Power Syst 31(5):3506–3515
13. Taylor JW, Mcsharry PE, Member S (2007) Short-term load forecasting methods : an evaluation based on European data. IEEE Trans Power Syst 22(4):2213–2219
14. Wu H, Shahidehpour M, Alabdulwahab A, Abusorrah A (2016) A game theoretic approach to risk-based optimal bidding strategies for electric vehicle aggregators in electricity markets with variable wind energy resources. IEEE Trans Sustain Energ 7(1):374–385
15. Pandit S, Kapur D (2015) Electric vehicles in India policies, opportunities and current scenario. ADB Open Innovation Forum, Manila, 20/05/2015. Available at: khub.niua.org/.../Electric_Vehicles__Policies_Opportunities_Scenario_1-SPandit.pdf
16. EY (2016) Electric vehicles adoption: potential impact in India. In: A Power and Utilities perspective, June 2016. Available at: http://www.ey.com -> EY-ev-adoption-potential-impact-in-India-july-2016.pdf19
17. Marathe SR (2013) xEV: Policies & implementation status in india. In: 2nd AAI—Summit, Electric Vehicle Session. Denpasar, Bali, Indonesia, 26th November 2013
18. Springer India—New Delhi Auto Tech Rev (2015) 4: 22. https://doi.org/10.1365/s40112-015-0912-y
19. Fame_India_Revised270415.pdf F. No. 21(37)/NAB/DIDM/2014

Part III
Renewable Energy, Microgrids and Energy Storage

Challenges in Implementation of Virtual Synchronous Generator

Ganesh N. Jadhav, Sadik J. Shaikh and Omkar N. Buwa

Abstract Grid-connected distributed energy resources (DERs) along with conventional power plants are the best solution to meet the increasing energy demand. But higher penetration level of these DERs may affect the system stability and its performance. DERs do not contribute any inertia unlike synchronous generators (SG) use conventionally. Absence of inertia results in large frequency variation, leading system to instability. To stabilize the system, virtual inertia can be added using short-term energy storage. This concept of virtual inertia is termed as virtual synchronous generator (VSG). This may be the basis for future grids with 100% renewable in India. The use of VSGs into the DERs interface to grid and off-grid is having several challenges. Electrical industry reliably integrates large amount of DGs/RESs into system; hence, it is important to pay attention for managing number of VSGs in the grid. The main aspect is to handle the changes in the system due to VSGs and to make the system robust, flexible and reliable. This article reviews the basic challenges in the implementation of VSG.

Keywords Distributed energy resources · Virtual synchronous generators
Renewable energy sources · Distributed generators · State of charge
Point of common coupling

G. N. Jadhav (✉) · S. J. Shaikh
Department of Electrical Engineering, K. K. Wagh. I. E. E. & R, Nashik,
Maharashtra, India
e-mail: ganyogesh81@gmail.com

S. J. Shaikh
e-mail: sjshaikh189@gmail.com

O. N. Buwa
Department of Marketing and Proposal, L & T Power, Vadodara, Gujarat, India
e-mail: onbuwa@gmail.com

1 Introduction

In present world scenario, electrical energy demand is rising phenomenally along with need of reliable, stable and quality power. But conventional power facing problems due to faster depletion of fossil fuel, poor energy efficiency and increased global warming. All this leads to a new trend of generating power at distribution level by using non-conventional energy sources. This type of power generation is called as distributed generation (DG). According to National Electricity Policy Draft of Central Electricity Authority, share of renewable till 31 March 2016 is 14% of total generating capacity. The targeted renewable energy generation is 20.3 and 24.2% of total energy requirement in 2021–2022 and 2026–2027, respectively [1].

Integration of DGs to power grid and realizing active distribution network require implementation of radially new concept, i.e. microgrid [2, 3]. Microgrid is a cluster of various microsources, energy storage devices and variable loads that operate as single control system. Microgrid works either in grid-connected mode or stand-alone mode [4].

Various disturbances in power system may lead to frequency deviation which can be reduced by adding amount of inertia in the system. This inertia is contributed by kinetic energy of rotating masses of synchronous generator and turbine. Inertia acts bidirectionally, i.e. it can be supplied to the system or absorbed from the system. Rotation less nature of DGs affects the inertia of the overall system. Reduced net inertia of the system severely affects stability and dynamic performance of the overall system [5, 6]. This problem of inertia can be overcome by providing deficient amount of inertia virtually in the system by a combination of short-term energy storage, power inverters and control mechanism. This concept is called as VSG [7].

Electrical industry reliably integrates large amount of DGs/RES into system. It is required to arrange number of VSG in the system for its operation. Important aspect is to handle the changes in the system, to make system flexible, reliable and robust. Implementation of VSG in actual practice has many challenges. This challenge may be either technical or non-technical. So, it is required study that challenges before implementing VSG for the better performance of the system.

This survey paper organized as follows: Sect. 1 discusses about basic concept of VSG, Sect. 2 explains advantages of VSG, Sect. 3 gives technical challenges in implementing VSG, Sect. 4 non-technical challenges in implementing VSG, and Sect. 5 gives conclusion of the paper.

2 Concept of VSG

Idea behind concept of VSG is combination of modern power electronic devices and grid supporting characteristics of electromechanical synchronous machine. Emulation of virtual rotational inertia in DG can reduce rate of change of frequency

(ROCOF). Injection or absorption of appropriate amount of power in/from grid takes place after sensing grid voltage/frequency and state of charge of storage device. Inverter is controlled to act as SG by VSG controller. VSG provides virtual inertia like rotational inertia of synchronous generator with help of storage devices, resulting the grid stability can improve. The inverter keeps continuous synchronization with the other generators in power system.

Equation of VSG from [7] can be given as:

$$P_{VSG} = P_0 + K_I \frac{d\Delta\omega}{dt} + K_P \Delta\omega \qquad (1)$$

where $\Delta\omega = \omega - \omega_0$ and ω_0 is the nominal frequency of the grid.

First term P_0 denotes the primary power that should be transferred to the inverter.

Second term indicates that the power will be generated or absorbed by VSG according to positive or negative initial rate of frequency change $\frac{d\Delta\omega}{dt}$

K_I is the inertia emulating characteristics given by Eq. (2);

where P_{g0} is the nominal apparent power of the generator and H shows amount of inertia

$$K_I = \frac{2HP_{g0}}{\omega_0} \qquad (2)$$

3 Advantages of VSG

Advantages of VSG can be summarized as below [9]:

- Unlike a real synchronous machine, the parameters of swing equation of VSG can be controlled in real time. This leads enhancement of the fast response of virtual machine in tracking steady state frequency.
- Challenges of SG like magnetic saturation and eddy current losses will be absent in VSG.
- VSG can determine proper control parameter of power system stabilizer(PSS) more easily.
- VSM has conceptual simplicity due to immediate and intuitive physical interpretation of its behaviour with analogy to corresponding behaviour of synchronous machine.
- VSG uses energy storage in its topology, when DERs operate parallel with main grid through VSM, surplus power in grid can flow towards VSG and thus can be stored in storage device. In conventional power system, reverse power flow will damage the prime mover due to monitoring of generator. In other words, it can be said that VSG can act as a motor.

4 Technical Challenges in Implementing VSG

The challenges in implementing VSG can be considered as technical and non-technical challenges as per the nature of the operating parameter. Following are some of the technical challenges while implementing VSG.

4.1 Measurement and Computational Techniques

Relation between power and frequency as per droop control can be given as:

$$\Delta P = K_{pf} * \Delta f \qquad (3)$$

where $1/K_{pf}$ is called as droop and given by R.

ΔP and Δf is variation in power and frequency from defined steady-state operating points.

Figure 1 indicates that VSG senses the change in frequency and responds according to the variation. Hence, frequency deviation detection plays an important role in VSG operation. Various methods are used for the frequency detection.

The first ever adopted method for the frequency detection is zero crossing method. In this method, exact half period time of the sinusoidal signal is calculated using data of two sequential zero crossing and linear interpolation for improving the detection accuracy. The frequency is calculated as $f = \frac{1}{2T_h}$. The corrected ZCD is very precise, but produces some wrong result in case of low-frequency voltage variation (e.g. flicker) and high-order harmonics (third, fifth, etc.). Use of filter

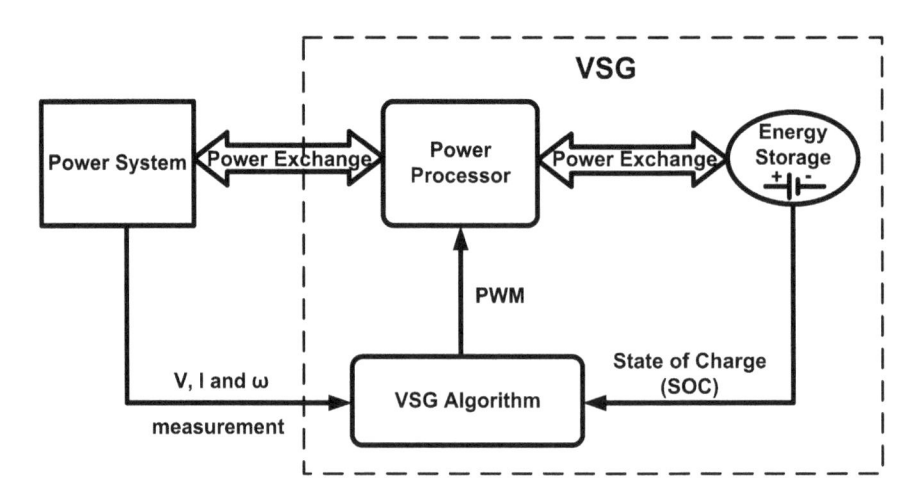

Fig. 1 Virtual synchronous generator [8]

before ZCD block allows filtering out unnecessary harmonics content but introduces certain delay [10].

The technique which is used to extract phase angle from grid voltage is called as phase locked loop (PLL) technique. This is a feedback control system, which adjusts the phase of locally generated signals automatically to match phase input signals. Most commonly used technique is SRF method, in which instantaneous phase angle is detected and synchronization of PLL rotating reference frame to the utility voltage vector takes place. The reference is locked to the utility voltage vector phase angle by means of a PI or PID controller which sets the direct axis reference voltage to zero. Figure 2 shows the conceptual diagram of the PLL.

Sometimes result from PLL is unsatisfactory as phase lead or lag may be introduced because of filter used in its structure to filter out voltage sensed. To overcome this, the advancement in conventional PLL was done. PLL is then used with PI and PID controller as shown in Fig. 3.

In first case, PI is used for locking the reference and real values by treating error detected. While in second case, three-phase reference signals are converted to the dq0 rotating frame using angular speed of internal oscillator. The quadrature axis of the signal, is proportional to the phase difference between the abc signal. The internal oscillator rotating frame is filtered with a mean block. The proportional integral derivative (PID) controller with an optional automatic gain control (AGC) keeps the phase difference to 0 by acting on controlled oscillator. The PID output, corresponding to the angular velocity, is filtered and converted to the frequency, in Hz, which is used by the mean value [11].

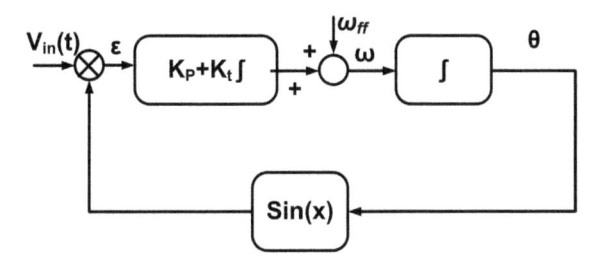

Research in the frequency detection technique introduces a digital method called as discrete Fourier transform (DFT). DFT-based frequency estimation algorithm has been used in many aspects of power system analysis. It has advantage like high precision, fast computation and strong immunity [12]. Problem with this method is DFT does not give accurate result at off-nominal frequency.

Fig. 2 Conceptual scheme of PLL [11]

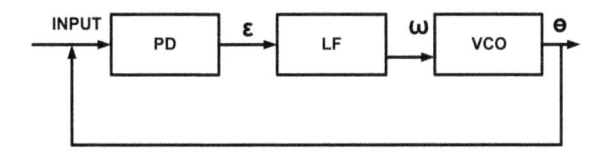

Fig. 3 Block diagram of the PLL [11]

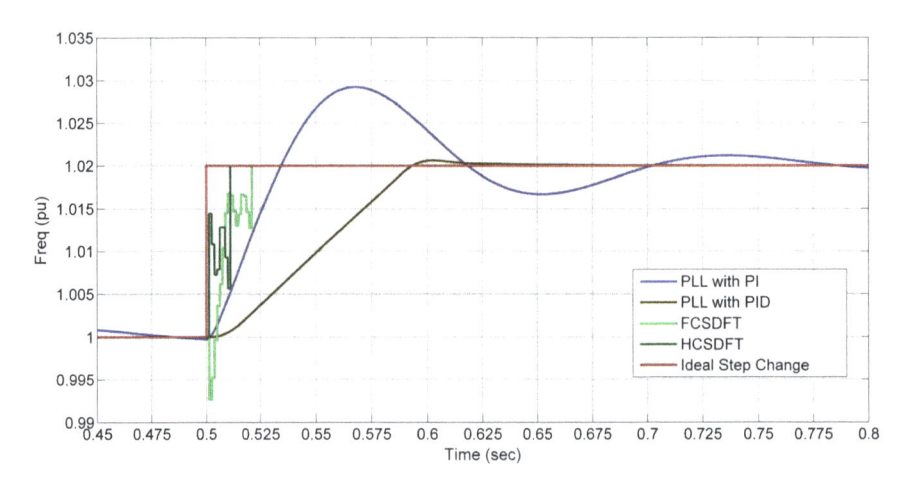

Fig. 4 Comparison of frequency detection techniques

Table 1 Comparison of different frequency detection techniques

Sr. no.	Frequency detection technique	Settling time (s)
1.	PLL with PI controller	0.4397 ($\approx$22 cycles)
2.	PLL with PID controller	0.1915 ($\approx$10 cycles)
3.	Full cycle SDFT controller	0.0213 ($\approx$1 cycles)
4.	Half cycle SDFT	0.0113 ($\approx$0.5 cycles)

Further advancement leads to smart discrete Fourier transform (SDFT) smartly avoids the errors that arise when frequency deviates from the fundamental frequency and keeps all the advantages of the DFT, e.g. immune to harmonics of fundamental frequency, obtaining easily the parameters of amplitude and phase and even the recursive computing can be used in SDFT.

Out of all the available for frequency detection, which method is to be utilized for particular system is challenge. This decision making depends upon the topology of the system, availability of inertia in system and delay required with reference to disturbances. The difference in the delays in different methods is shown in Fig. 4, and detailed analysis is given in Table 1.

4.2 Modelling and Analysis Tools

VSG can be modelled for emulating characteristics of either stator or rotor or both stator and rotor of SG. Detail modelling of VSG according to requirement of the system is explained in [13].

Response of SG to grid disturbances depends upon the order of modelling of SG. In the literature, SG modelling reference is available for higher order like sixth or seventh. But modelling of VSG has been done using first- or second-order. Equation for first-order and second-order VSG given in [14] as:

$$J\frac{d\omega_m}{dt} = T_m - T_e = T_a \quad (4)$$

$$J\frac{d^2\delta}{dt^2} = T_m - T_e - K_D\Delta\omega_m \quad (5)$$

where

J is the combined inertia of turbine and generator,
ω_m is angular speed of the rotor,
T_a is accelerating torque,
T_m is mechanical torque,
T_e is the electromagnetic torque,
K_D represents the damping factor,
δ represents angular position of the rotor.

Grid may require active power as well as reactive power at the same time. Hence, to control both at same time is one of the challenges while designing VSG. Instantaneous output power of VSG pulsates twice the line frequency ($2f_{line}$) if local loads or voltage is unbalanced. This leads to double line frequency ripple (DLFR) in frequency and amplitude of output voltage of VSG, which is then harmful to critical load. Because of this, the bandwidth of the power of VSG is less than ($2f_{line}$). Root locus method is used for designing of the VSG parameter, but the consideration of attenuation requirement of DLFR is not done, thus designed parameters are not good. However, due to the coupling effect between the active power loop and reactive power loop, the parameter design is very difficult, and control parameter is designed by trial and error basis [15].

As VSG basic operation depends upon the frequency signal, a number of methods are available for detecting frequency and supplying the power in the loop. In [7, 9, 16–18], explains the different topologies/modelling of the VSG, but the aim of all the system remains same.

4.3 Effective Algorithm Development

Important components of VSG system are power processor, energy storage device and algorithm which regulates that exchange power between energy storage and power system. Frequency variation initiates power exchange from storage system to grid. Based on the share of power, it classified into two types as:

1. *Linear VSG Algorithm:* As explained in Sect. 1, power contribution totally depends upon system frequency variation as given in Eq. (6) which is known as "Linear VSG Algorithm"

$$P_{VSG} = K_i \frac{d\omega}{dt} \qquad (6)$$

where

P_{VSG} is output of VSG,
$\frac{d\omega}{dt}$ is input signal,
ω is synchronous speed.

The value of K_i is chosen, so as the VSG will exchange its nominal power for a specific value of $\frac{d\omega}{dt}$. It indicates, for small disturbances, the VSG effect is also small and for large disturbances the power share is large. In any case, P_{VSG} have specific duration and shape and energy injected E_{VSG} can be calculated as [19]:

$$E_{VSG} = \int_{t_0}^{t_1} P_{VSG}(t) dt \qquad (7)$$

VSG operates during transition from one steady state of frequency f_0 to another steady state f_1. The time difference from Eq. (7) is $t_0 - t_1 = t_{trans}$, i.e. duration of transient. Hence, energy of VSG (E_{VSG}) is the work done Wd_{VSG}, during transition time:

$$Wd_{VSG} = \frac{1}{2} J_V \omega_0^2 - \frac{1}{2} J_V \omega_1^2 \qquad (8)$$

Rearranging Eq. (8), we get the equation of virtual inertia as:

$$J_V = \frac{2 E_{VSG}}{(\omega_0^2 - \omega_1^2)} \qquad (9)$$

Equation (9) states that for constant energy (E_{VSG}), J will vary according to size of disturbance. Power supply by VSG will be according to the requirement of the power requirement of the circuit [19]. It is observed that as per the demand of grid, the power dispatch from VSG takes place.

2. *Binary VSG Algorithm:* Stability studies in islanded system show machine with fast spinning reserve is beneficial for frequency stability. Also by increasing K_I for constant input power, better frequency oscillation damping is achieved. This happens because VSG makes power available in same time. VSG is power

electronic-based device, which has specific rating, i.e. supply power limiting to −1 pu < Pi < +1 pu. Virtual inertia becomes more effective, when more power is made available faster.

To complete this requirement, new algorithm is derived for emulating rotational inertia. This algorithm is called as "Binary VSG algorithm". In grid operation, ROCOF relay acquires $\frac{d\omega}{dt}$ from power system, compares it with reference value, and forwards trip signal to the relay. Similarly, this technique is used for emulating inertia. Instead of $\frac{d\omega}{dt}$ serving signal to trip a relay, it is used to trigger VSG to give maximum power. The equation of this can be given as:

$$P_{VSG} = \text{PD} \times \text{sw} \times P_{VSG\_MAX} \qquad (10)$$

where PD is power direction and Sw stands for switch either 0 or 1 [19].

As observed in [19] as grid demand crosses upper/lower boundaries, the power dispatch from VSG takes place. This power dispatch is either zero or 1, i.e. no power or full power.

For using VSG in the actual grid, it is very much important to consider all parameters such as parameters of the equipment's used, capacity of the storage system, performance of the algorithm on frequency change. All these parameters should be considered during modelling and simulation of VSG.

There are several software tools available for simulation purpose like MATLAB, PSCAD, DIgSILENT power factory which are well used. Amongst all the tools, selection of the appropriate tool is a challenging part. Results from each tool are different. Modelling efficiencies also have the different nature.

4.4 Coordination Between VSG and SG

In introduction, the importance of the inertia in power system is described. SG supplies require inertia in case of the normal power system, by increasing or reducing the speed of rotor by governor action. The same inertia is provided by VSG with energy storage as per set algorithm and detected ROCOF. At same time SG also responds to the system behaviour.

Important features of VSG are rapid injection/absorption of active power to/from grid. Rapid interaction of active power with grid causes net grid power imbalance, as the active power given by VSGs changes suddenly to recover the system frequency followed by disturbances. Increased/decreased power in grid can last just for few seconds. Conventional SGs operate for changed demand by shifting generation to compensate power imbalance in grid, but fast power injection by VSGs may slow down the response of conventional SG.

The need of more spinning reserve in the system is explained in [19]. It is also indicated, more is the spinning reserve, more is the power which can be supplied to the main grid by VSG. Model of VSG explains the energy storage device is having great importance in VSG topology. Power exchange with the grid totally depends upon the energy storage and control algorithm (Fig. 5). This battery algorithm structure is mainly divided into two parts [20]:

- The capacity model, which estimates the state of charge (SOC) of batteries.
- The voltage model that calculates the terminal voltage of batteries.

The capacity model can address to both recovery and rate capacity effect of the battery. First refers to effect where amount of charge becomes available to battery when no charge current is presented, and second refers when less charge is drawn from battery and discharge current is increased.

In VSG, the power electronics inverter is used, whereas in SG, mechanical system is used. Though the parameters of both are same, the behaviour of SG and VSG is different. Rapid power injection by VSG in the grid creates frequency oscillations as shown in Fig. 6.

4.5 Optimal Location of VSG

Literature of VSG indicates that placement of VSG is in between point of common coupling (PCC) and microgrid. No reference standards are available for this placement/topology. If location of VSG varies, behaviour/response will change according to position. VSG can be placed near to the microgrid unit having DC generation or beyond PCC. Each of this location of VSG has different behaviour/response. In a nutshell, optimal placement of VSG in a given system is challenge.

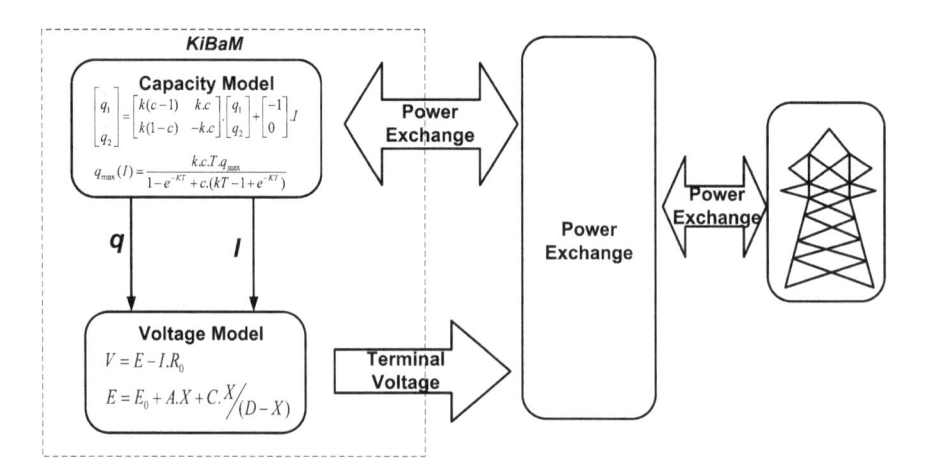

Fig. 5 Principle of kinetic battery model [20]

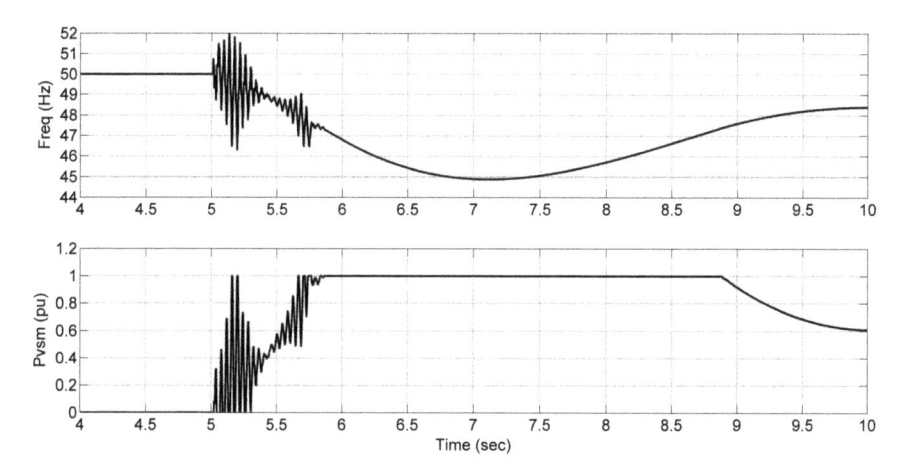

Fig. 6 Frequency response during rapid power injection by VSG

5 Non-technical Challenges

Following may be the some of non-technical challenges.

5.1 R and D in Energy Storage System

As discussed in above sections, VSG functionality totally depends upon the storage system and power sharing algorithm. Hence, it is important that system should be present with the plenty of spinning reserve, considering SOC effect also. Rapid supply of power should be available to damp out the oscillations in frequency; hence, instant discharge of power should happen. To withstand with all the statement listed above, it is very much important to do the research work in storage system. Storage device should be available with enough spinning reserve and well-arranged algorithm. Optimum investment and R & D efforts are required for success of this technology.

5.2 Cost Effectiveness

As discussed in above sections, important components of VSG system are power processor, energy storage device, and algorithm which regulates that exchange power between energy storage and power system. Better operation can be obtained by updating the existing technology with the new inventions. While updating the new technology for all equipment's, the cost required will be higher. Also, storage system used in the VSG should be available with huge power reserve; hence, capacity of the storage system should be large.

For instant availability of power to grid, continuous research is required; hence, the cost of overall system is high. So while building the VSG, cost factor should be considered.

5.3 Economical Load Dispatch

Economical load dispatch and unit commitment problem are very difficult to handle in the presence of DER. When inverter-based DER is introduced as a VSG, it is difficult to solve the economical load dispatch problem for power pools. As the response of VSG depends on frequency variations, power variations are frequent in the system.

6 Conclusion

Inclusion of DER/RES in the grid reduces the net inertia of the system. Hence, it is required to add the inertia virtually in the system by means of VSG. But implementation of VSG in actual has some of the challenges which must be considered while building the system. VSG depends on frequency variation; hence, it is required to adopt better method for frequency detection. Similarly, the challenges of modelling and algorithm development should be taken under consideration. VSG and SG coordination is also important as both acts on the same time. Also, non-technical challenges, cost and R & D in storage system should be considered which not takes part in the system but hamper system performance externally.

References

1. Government of India Ministry of Power Central Electricity Authority (2006) Draft national electricity plan, Dec 2016
2. Lasseter RH, Paigi P (2004) Microgrid: a conceptual solution in power electronics specialists conference, PESC 04. In: IEEE 35th Annual, vol 6, June 2004, pp 4285–4290
3. Wang B, Sun M, Dong B (2011) The existed problems and possible solutions of distributed generation microgrid operation. In: Power and energy engineering conference (APPEEC), Asia-Pacific, Mar 2011, pp 1–4
4. Zhang H, Li S (2011) Research on microgrid. In: International conference on advanced power system automation and protection (APAP), vol 1, Oct 2011, pp 595–598
5. Ulbig A, Borsche TS, Andersson G (2013) Impact of low rotational inertia on power system stability and operation. ArXiv e-prints
6. Abreu LVL, Shahidehpour M (2006) Wind energy and power system inertia. In: 2006 IEEE power engineering society general meeting, p 6

7. Bevrani H, Ise T, Miura Y (2014) Virtual synchronous generators: a survey and new perspectives. Int J Electr Power Energy Syst 54:244–254
8. Karapanos V, de Haan S, Zwetsloot K (2011) Real time simulation of a power system with vsg hardware in the loop. In: IECON 2011 37th annual conference of the IEEE industrial electronics society, Nov 2011, pp 3748–3754
9. Yu M, Dyko A, Roscoe A, Booth C, Ierna R, Urdal H, Zhu J (2015) Effects of swing equation based inertial response (sebir) control on penetration limits of nonsynchronous generation in the gb power system. In: International conference on renewable power generation (RPG 2015), Oct 2015, pp 1–6
10. Albu M, Calin M, Federenciuc D, Diaz J (2011) The measurement layer of the virtual synchronous generator operation in the field test. In: IEEE international workshop on applied measurements for power systems (AMPS), Sept 2011, pp 85–89
11. Nicastri A, Nagliero A (2010) Comparison and evaluation of the pll techniques for the design of the grid-connected inverter systems. In: IEEE international symposium on industrial electronics, July 2010, pp 3865–3870
12. Buwa ON, Jadhav GN (2016) A dissertation report on virtual synchronous generator
13. Arricibita D, Sanchis P, Marroyo L (2016) Virtual synchronous generators classification and common trends. In: IECON 2016 42nd annual conference of the IEEE industrial electronics society, Oct 2016, pp 2433–2438
14. Xiong L, Liu X, Wang F, Zhuo F (2016) Static synchronous generator model for investigating dynamic behaviors and stability issues of grid tied inverters. In: IEEE applied power electronics conference and exposition (APEC), Mar 2016, pp 2742–2747
15. Du Y, Guerrero JM, Chang L, Su J, Mao M (2013) Modeling, analysis, and design of a frequency droop based virtual synchronous generator for microgrid applications. In: IEEE ECCE Asia Downunder, June 2013, pp 643–649
16. Ashabani M, Freijedo FD, Golestan S, Guerrero JM (2016) Inducverters: PLL-less converters with auto synchronization and emulated inertia capability. IEEE Trans Smart Grid 7(3):1660–1674
17. Chen Y, Hesse R, Turschner D, Beck HP (2011) Improving the grid power quality using virtual synchronous machines. In: International conference on power engineering, energy and electrical drives, May 2011, pp 1–6
18. van Wesenbeeck MPN, de Haan SWH, Varela P, Visscher K (2009) Grid tied converter with virtual kinetic storage. In: IEEE Bucharest PowerTech, June 2009, pp 1–7
19. Karapanos V, Kotsampopoulos P, Hatziargyriou N (2015) Performance of the linear and binary algorithm of virtual synchronous generators for the emulation of rotational inertia. Electr Power Syst Res 123:119–127. [Online]. Available://www.sciencedirect.com/science/article/pii/S0378779615000449
20. Vassilakis A, Kotsampopoulos P, Hatziargyriou N, Karapanos V (2013) A battery energy storage based virtual synchronous generator. In: IREP symposium bulk power system dynamics and control—IX optimization, security and control of the emerging power grid, Aug 2013, pp 1–6

Efficiency Gain Using DC Microgrid and BLDC Machine-Based 48 V Air Cooler

Sriram Narayanamurthy, Pradheep Ganesan, Ashok Jhunjhunwala and Prabhjot Kaur

Abstract A desert cooler typically is a blower fan that sucks air through a moistened surface to achieve a drop in inlet air temperature. A pump is used to circulate water from a lower tank to the evaporative pads working with a swing louver motor for directing the air automatically. The power consumption of the evaporative air cooler is primarily in single phase induction motor (SPIM) used for fan motor [1, 2]. To lower this power consumption, a brushless DC machine and its circuitry were designed and used. This is powered also by PV panel and batteries. The power consumption comparative study was carried out on SPIM and BLDC machines at three speeds. The outcome of the study shows 39.3% of power saved on with BLDC machine air coolers.

Keywords BLDC machine and its circuitry · Power savings · Evaporative cooling pad

---

S. Narayanamurthy (✉) · P. Ganesan · A. Jhunjhunwala
Electrical Engineering Department, Indian Institute of Technology Madras, Chennai, India
e-mail: sriramn@tenet.res.in

P. Ganesan
e-mail: pradheepganesan@hotmail.co.uk

A. Jhunjhunwala
e-mail: ashok@tenet.res.in

P. Kaur
Reliance IITM Telecom Centre of Excellence, IIT-Research Park Phase-1, Chennai, India
e-mail: prabhjot@tenet.res.in

© Springer Nature Singapore Pte Ltd. 2018
R. K. Pillai et al. (eds.), *ISGW 2017: Compendium of Technical Papers*, Lecture Notes in Electrical Engineering 487, https://doi.org/10.1007/978-981-10-8249-8_19

1 Introduction

INDIA is a country with 54% of the households living in hot and dry climates facing thermal discomfort which is non-conducive for various everyday human activities.

Air coolers in INDIA have a market size of 4000 crores with an annual growth rate of 11%. Any impact on the power consumption would lead to a major change in the power generation requirement and lesser carbon emissions.

About 50 million homes are not connected to grid supply, while 100 million of them have load shedding between 2 and 12 h a day. This calls for looking at options of alternate energy sources and storage technologies for enabling operation during outages.

While 50% of Indian homes cannot afford power at subsidized rates of 5 rupees per unit, a solution of DC microgrid is an innovative idea to solve the challenges [6].

Evaporative cooling is a technology which uses the concept of evaporating water and adding it to the incoming airstream. This method changes the composition of the outgoing air while taking the heat for evaporation from the air.

The simplicity of the technology and its application in wide areas gives us the motivation to explore avenues to reduce the power consumption and make a significant difference to the overall power demand.

The efficiency of the air cooler is decided by the amount of water that the pad can evaporate for a given flow of air over it. The lowest temperature of air that a direct evaporative cooler can deliver is the wet bulb temperature.

One of the other aspects that decide the efficiency of the air cooler is the primary mover for the fan blower. The motor consumes the maximum power of all the consuming components including the pump and swing motor. The objective is to employ a motor that is more efficient than the conventionally used induction motor and be able to integrate to renewable sources without much of technical complexity and ownership costs.

2 Methodology and System Specification

The main moving part of the air cooler being the fan blower motor is usually a single phase induction machine (SPIM). The power consumption of SPIM and a custom-designed brushless DC (BLDC) motor is compared by running the motor at three different RPM such as 1395, 1350, and 1280 with 16" metal fan blade. The flow rates and power consumption of AC and DC pumps were also measured to select the optimum flow rate. Finally, the motor, pump, and evaporative pads were integrated on to the cooler as shown in Fig. 1.

To calculate the efficiency of cooling pad, thermocouples and humidity sensors were installed on all the sides of the cooler to log the air-in and air-out temperature

Fig. 1 Experimental setup

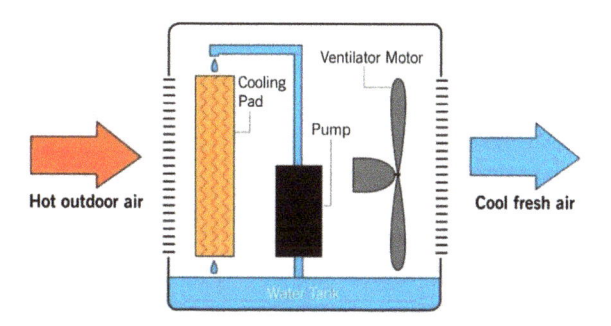

Table 1 Specification of the air cooler

Operating voltage	40–60 V DC
Fan blade diameter	14″
Fan blade RPM	1100–1350 RPM
Motor power	20–70 W
Pump power	5 W
Swing motor power	5 W
Water tank capacity	50 L
Cooling media	Honey comb pad/Wood wool pad
Product dimensions (mm)	558*596*940 (L*W*H)

at different relative humidity (RH) values. The energy consumption of the cooler is logged though power meter (Table 1).

Specification of the retrofitted desert cooler system is as follows:

3 System Operation and Governing Equations

Initially, water is completely filled in the cooler's bottom tank and circulated to the cooling pads using a lift pump. The blower is switched ON to extract the unsaturated air from the surroundings and passed through the evaporative pads. As the incoming airstream passes through the surface of the wet pads, the water evaporates into it resulting in increase of moisture content in the air. The wet bulb temperature of air represents the lowest temperature that the incoming air can reach only using evaporation of water. Once the system runs for a certain period, the incoming air will have a higher humidity in a closed room scenario and will reach closer to the wet bulb temperature [3–5].

Air will never reach the wet bulb temperature, as some of the air will pass without getting in contact with the incoming stream of air.

4 Drive Circuitry Description and Architecture

A motor drive was designed and built according to the specification of the compressor motor using Infineon-based solution with the architecture of a three-legged inverter operating out of a 48 V DC bus as shown in Fig. 2.

The drive circuit contains various features of current sensing, over voltage, and over current protection as shown in Fig. 3. A closed loop operation of the drive ensures that the RPM of the fan motor is maintained at the desired speed.

5 Result and Discussion

Figures 4, 5, and 6 represent power consumption of the system components. Power consumption of the BLDC motor at highest fan speed is only 44% of SPIM machine. The BLDC pump delivers almost equal flow rate on comparing with SPIM by consuming only 9.5 W. Finally, BLDC swing motors save 45% of power. These small auxiliaries of the system save enormous energy utilization compared to SPIM system and represented in Fig. 7.

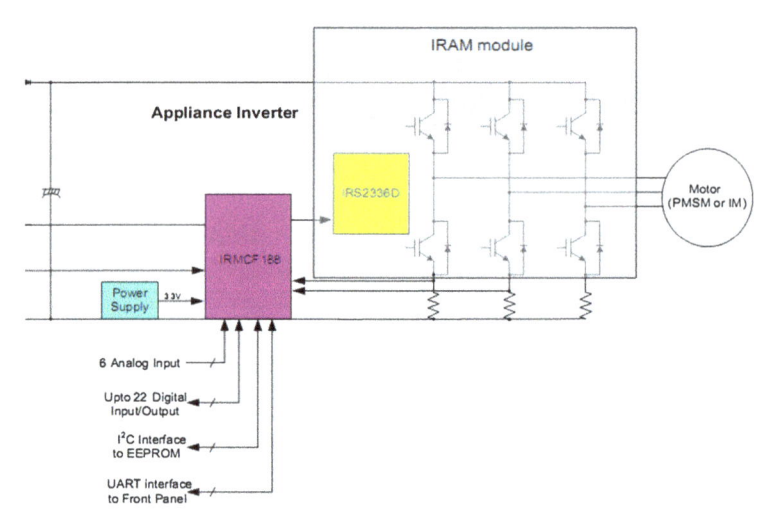

Fig. 2 System architecture

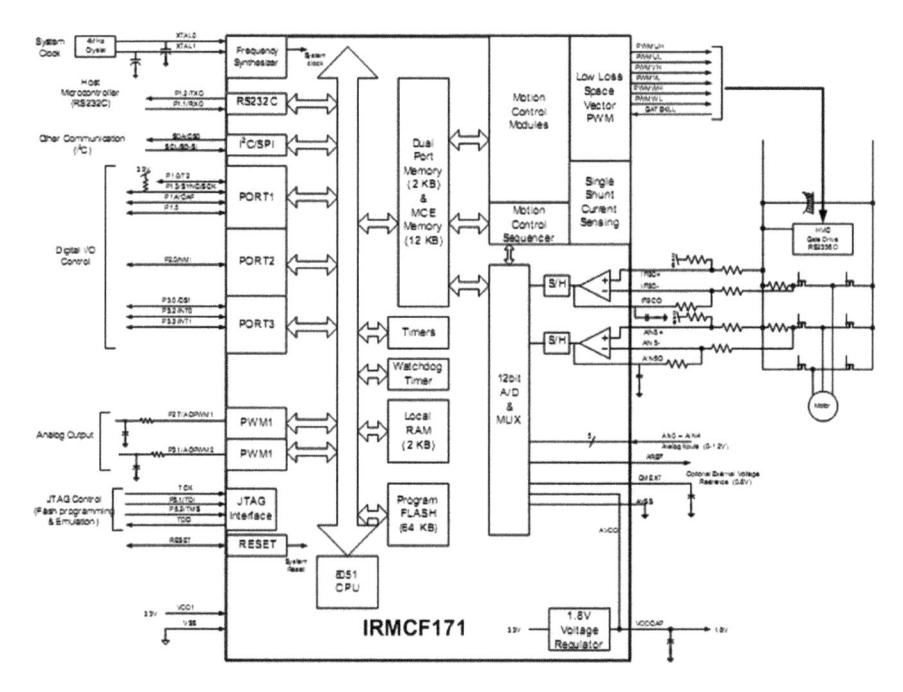

Efficiency Gain Using DC Microgrid and BLDC Machine-Based ...   227

Fig. 3 Application architecture

MOTOR	SPEED RPM	VOLTAGE (V)	CURRENT(A)	POWER FACTOR	POWER (W)
SPIM	1395	228	0.69	0.99	156
	1350	228	0.61	0.99	138
	1280	228	0.56	0.98	125
BLDC	1395	48	1.82	1	87
	1350	48	1.67	1	80
	1280	48	1.47	1	71

Fig. 4 Fan motor test results

PUMP	VOLTAGE(A)	CURRENT (A)	POWER FACTOR	POWER (W)	FLOWRATE(kg/h)
SPIM	229	0.11	0.5	13.4	480
	6	0.6	1	3.6	334
BLDC	8	0.8	1	6.4	393
	10	0.95	1	9.5	453
	12	1.12	1	13.6	511

Fig. 5 Pump test results

SWING MOTOR	VOLTAGE(V)	CURRENT (A)	POWER FACTOR	POWER (W)
SPIM	229	0.014	0.872	2.75
BLDC	15	0.1	1	1.5

Fig. 6 Swing motor power comparison

COOLER TYPE	SPEED (RPM)	TOTAL POWER (W)	AVERAGE POWER SAVING %
SPIM COOLER	1395	172.15	39.3
	1350	154.15	
	1280	141.15	
BLDC COOLER	1395	102.1	
	1350	95.1	
	1280	86.1	

Fig. 7 Total power savings

6 Conclusion

Retrofitting a BLDC motor instead of a SPIM for the fan blower, pump, and swing motor saves about 39.3% power consumption. The 48 V DC controller also opens up avenues for using the cooler in power deficit areas and in locations where there is no power availability completely, due to the operating current being the same as starting current.

References

1. ASHRAE (1988) ASHRAE Handbook–Equipment, Chapter 4, pp 41–43
2. ASHRAE (1989) ASHRAE Handbook–Fundamentals, Chapter 1, pp 26–28
3. Stoecker WF (1971) Design of thermal systems. McGraw-Hill, New York
4. Watt JR, Crow LW, Greenberg A (1986) Evaporative air conditioning handbook. Chapman & Hall, New York
5. Anderson WM (1986) Three-stage evaporative air conditioning versus conventional mechanical refrigeration. ASHRAE Trans 92(1B):358–370
6. Jhunjhunwala A, Lolla A, Kaur P (2016) Solar-dc microgrid for Indian homes: a transforming power scenario. IEEE Electrification Mag 4:10–19

Integrating Energy Efficiency with Renewables and Energy Storage for a Smarter and Greener Residential Solution

Satish Kumar, Smita Chandiwala, Vaibhav Rai Khare and Manish Pant

Abstract India's solar rooftop policy is a big boost for distributed generation of electricity. It is also an opportunity to develop an integrated energy policy that will bring about a market transformation by closely coupling distributed energy generation and existing energy storage solutions. If executed properly, this kind of integrated energy policy can help to optimize the systemic cost of supplying power to both on-grid and off-grid Indian homes. At present, lead-acid cells in inverter systems are the most recognizable battery system and have a significant installed base as the Indian home inverter market has seen a CAGR growth of 20% in the last 7–8 years. This paper will investigate a standard, cost-effective solution for retrofit of existing inverter-battery installations and rooftop solar power for residential power needs based on simple payback calculations. This will include a feasibility analysis entailing load calculations and financial viability of the solar retrofit solution for Indian power market scenario. The paper will also estimate the total benefits to the nation from a large-scale roll-out of this solution, from the perspective of capacity generation and reliability improvement.

Keywords Solar rooftop policy · Solar retrofit solution · Inverter-battery systems

---

S. Kumar (✉) · V. R. Khare
Alliance for an Energy Efficient Economy, New Delhi, India
e-mail: chairman@aeee.in

V. R. Khare
e-mail: vaibhav@aeee.in

S. Chandiwala
Energe-se | Building Energy Narratives, New Delhi, India
e-mail: smita@energese.in

M. Pant
Schneider Electric India Pvt. Ltd., New Delhi, India
e-mail: manish.pant@schneider-electric.com

1 Introduction

India faced peak demand power shortages of 3.2% and energy shortages of 2.1% in 2015–16 (CEA) in spite of sufficient installed generation capacity due to multiple reasons such as inadequate coal supply, high distribution and transmission losses, the poor financial health of utilities and their inability to hence pay for expensive peak power purchases.

This has resulted in power demand-supply mismatches on a daily basis and utilities often opting for load shedding through power cuts. As a result, over the years, there has been a significant installed base of lead-acid inverter-battery storage systems to provide for critical backup loads in residential buildings as well as small retail businesses. Though the trend for peak demand-supply gap has been steadily decreasing, especially in big cities, in smaller cities and towns, frequent power cuts persist leading a steady growth in installation of these systems.

There is also an increasing push towards renewable energy technologies such as solar photovoltaic (PV) due to energy access, energy security, climate change mitigation and other environmental benefits. Solar PV systems provide a clean alternative to meeting power demands in these units with existing battery storage solutions, while reducing the consumption of the grid drawn electricity that is used for charging these systems.

This paper will investigate a standard, cost-effective solution for retrofit of existing inverter-battery installations and rooftop solar power for use as a backup for critical loads. The paper's objective is to provide an evaluative understanding of the benefits of this approach, the opportunity cost and its feasibility in terms of simple payback.

2 The Opportunity

According to *Electronics for You* magazine article on *Indian UPS and Industry* report, the Indian home inverter market is likely to see a CAGR growth of 20% from 2013 to 2020. At present, lead-acid cells in inverter systems are the most recognizable battery system with a significant installed base and growing steadily. These systems continuously draw power from the grid to store charge in the batteries for use during power cuts. Using a retrofit solution, combining solar photovoltaic panels to charge the existing battery storage capacity provides benefits of 'freeing' up utility power that is needed to charge batteries for other uses while providing a cost-effective solution to the inverter owners.

3 Unit Level: Technology Details, Costs and Payback

Typically, in a 2–3 BHK house, load requirement is around 600–700 W for running 3–4 fans, 3–5 lights (tube lights, CFLs, LED lamps), and a television during power outages.

A typical 900 VA inverter capable of meeting 720 W of load and fitted with a 150Ah battery, would require 4 × 100 W, 12 V PV panels to effectively provide electricity backup and a solar retrofit solution (SRS) to enable intelligent charging along with supporting wires, structures and installation. The total cost break-up for such a system amounting to INR 24,000 is given below (Source: Luminous Power Technologies).

Additionally, once the batteries are charged, the SRS ensures that the solar power can be fed through the inverter into the inverter circuit in homes; hence, utilizing the surplus solar energy optimally for meeting the household needs at the point of generation, leading to further savings of grid-based electricity. Hence, all solar power produced is used at the household level. This is in contrast to the current rooftop solar PV policy, which is encouraging the use of the grid as a battery bank to feed in surplus power and draw electricity from, when additional electricity is needed combined with net metering (Table 1).

This is a win-win interim solution for utilities facing issues in operationalizing net metering policies as well as technical issues faced by manufacturers in providing adequate security and reliability provisions for 'anti-islanding' technology in the design of inverters.

This system is also efficient as PV produced DC power is used directly to charge batteries, avoiding the conversion losses to the tune of 15% that occur in grid-based AC power for the same.

At a single unit level, the panels will pay for themselves in 6–7 years depending on the electricity cost (assumed to be within the range of INR 5–7/kWh), while the consumer will benefit from the additional electricity savings throughout the 25-year life of the panels. Over the years, batteries can be replaced if required, users can also add more panels and upgrade inverters to support increasing load requirements or connect to the grid (Table 2).

Table 1 Cost break-up for a solar retrofit solution

Part	Specifications	Assumptions	Costs (INR)
Panel	400 W, 12 V	50 Rs/Wp, 100 W–12 V 4 panels	20,000
Inverter	850–900 VA	*Already installed*	
Battery	150 Ah	*Already installed*	
Solar retrofit solution	12 V, 20 A	Supports 20 A charging	2000
Installation, structures, wires			2000
Total			24,000

Table 2 Payback calculation for solar panels

Key parameter	Numbers	Units or assumptions
Panel capacity	400	W
Panel efficiency	85%	Efficiency
Capacity utilization	21%	(Assuming 5–6 h of sun every day)
Inverter efficiency	90%	
First-year solar generation	563	kWh
*Electricity cost	5–7	INR/unit
Escalation in electricity costs	7%	per annum
Derating in Solar panels	0.5%	per annum
First-year electricity savings	2815–3940	INR
Payback (*range related to the electricity cost)		
Simple payback	6–7 years	
Internal rate of return	17–22%	
Net present value	18,576–34,734	INR

4 National Level: Cost and Avoided Energy Generation

According to National Sample Survey (2012) result and extrapolated to current year, the approximate installed base of inverters is around 25 million units. Assuming the above solution, if all of this capacity is retrofitted, it provides a 10 GW opportunity or an avoided grid-based capacity equivalent to 12.5 GW, assuming 20% aggregate technical and commercial losses. The total installation cost is estimated to be INR 60,000 crores with an effective cost of INR 4.8 Cr/MW. This is comparable to the costs of coal-based plants (3–7 Cr/MW), Solar PV power projects (4–8 Cr/MW) and lower than Hydro Power (6–10 Cr/MW). Moreover, Operations & Maintenance (O&M) costs are going to be miniscule as compared to the O&M expenditure for the coal-based thermal plants (Table 3).

Table 3 Cost of installation and effective cost

Key parameter	Numbers	Units or assumptions	
Approximate installed base of inverters	25.00	Million units	
Average inverter rating	900	VA	
Connectable to panel of rating	400	W	
Technology required	Solar panel 400 W + Solar Retrofit 12 V 20 A		
Cost of this technology	24,000	INR	
Total market size	10	Gigawatts	
Equivalent centralized power	12.5	GW	(20% AT&C losses)
Cost of installation	60,000	Cr	
Effective cost per MW	4.8	Cr/MW	

5 Conclusion

The solar retrofit solution (SRS) for existing inverter systems provides a viable solution for optimally utilizing the extensive existing lead-acid battery storage capacity in the country, while effectively reducing the dependence on grid-based electricity for maintaining electricity backup for critical loads for households and small retailers.

Government policies are actively supporting development of 40 GW of on-grid rooftop Solar PV though appropriate mechanisms such as financial subsidies, accelerated depreciation, net metering. These are based on envisioning the grid as a battery bank where surplus electricity produced by rooftop solar PVs can be fed in and drawn from the grid, as per requirements. This requires a smarter grid with sophisticated demand-supply balancing and capabilities of handling larger amounts of solar PV generated electricity allowing for two-way flow of electricity.

While this is the long-term solution for shifting towards cleaner and renewable sources of electricity supply, the proposed solar retrofit solution provides a cost-effective, efficient and quick transition to increasing rooftop PV consumption at the point of use first; especially for residential consumers which have seen the least adoption rates of solar PV in comparison to commercial and industrial consumers.

Seen within this perspective, there is a case to be made for including the solar retrofit solution within the 40 GW target. This could boost the solar rooftop PV adoption at a faster rate and at a substantially lesser cost for consumers, while continuing to meet their needs of backup power.

It will also be useful to determine how these could be integrated as either grid-connected systems (without storage) or as hybrid systems (grid connected as well as continue to provide backup power in the case of power failure) in the long term.

Reference

1. Central Electricity Authority (CEA), Load generation balance report 2016–17. http://www.cea.nic.in/reports/annual/lgbr/lgbr-2016.pdf

Smart Solutions and Opportunities for Key Challenges in Renewables Integration and Electric Vehicles Integration to a Conventional Grid

Goutham Yelmanchli

Abstract Internationally developed, developing, and underdeveloped countries are aiming toward energy trilemma (energy security, environmental sustainability, and energy equity). This can be made possible with the integration of renewable energy (RE) sources (mostly solar and wind). However, Reintegration with conventional grids brings with it grid stability issues, unpredictable and unbalanced supply and demand curves, reverse power flows ultimately resulting in blackouts as conventional grids are not ready to handle Reintegration. This is due to intermittent, variable, non-predictable, and location-specific generation profiles of renewables. All these challenges can be addressed using smart grid solutions integrating information and communication technology infrastructure with electricity infrastructure and adding intelligence to the combination. Smart technologies that can be readily implementable to overcome the above-addressed challenges are detailed in the paper. Some of these are demand response with customer engagement through smart meters, prosumers—encouraging customer feeding back to grid during peak demand through rooftop renewables, RE microgrids to feed a particular boundary isolated from conventional grid, smart inverters which act as storage devices and can feed AC back to grid, integrated storage using battery electric vehicles (BEVs), dynamic line rating using PMUs for real-time and enhanced loading and protection of feeders, etc. Benefits added to the utility and consumers of these technologies are discussed in detail. Electric mobility of transport is another key focus area worldwide due to depleting primary fuels and environmental (GHG) concerns. Major challenge to a distribution utility is the increased demand of plug-in electric vehicles (PEVs) and the charging infrastructure required for EVs. Smart solutions result in new business opportunities for distribution utilities using EV charging infrastructure and also benefit the customer through offering credit for feeding back to grid during peak hours and emergency islanding conditions. This is possible through smart integration of EV with conventional grid through two-way (electrical and information) communication between a distribution utility and consumer.

G. Yelmanchli (✉)
Distribution Customer Acquisition, The Tata Power Company Limited, Mumbai, India
e-mail: gcyelmanchli@tatapower.com

This paper highlights the challenges posed by the emergent requirement of EVs and smart grid solutions to overcome the hurdles of peak demand, renewable uncertainty, etc.

Keywords Smart · Grid · xEV · BEV · ISGF · Microgrids · PMU Renewable integration

1 Energy Trilemma

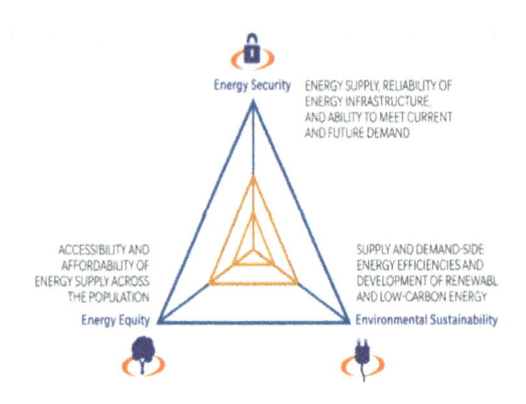

1.1 Introduction to a Smart Grid:

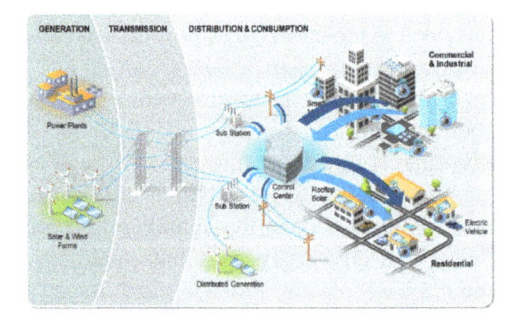

It is an electrical grid that uses ICT together and acts on information in an intelligent way to improve the efficiency, reliability, economics, and sustainability of the power supply. The smart grid facilitates two-way communications between suppliers and consumers and permits both the suppliers and the consumers to be more flexible and responsive in their operational strategies.

2 Reasons for Its Emergence

- Integrate more intermittent renewables such as wind and solar
- Integrate more distributed energy resources such as rooftop solar and electric vehicles
- Adhere to environmental mandates and constraints
- Balance limited energy supply and increasing peak demand
- Modernize aging energy infrastructure
- Distribution networks that can both "deliver" and "exchange" energy
- More engaged and empowered consumers
- Modernized energy infrastructure with operational technology (OT) and information technology (IT) integration
- Operating resiliently against various hazards.

3 Key Smart Grid Segments

(a) AMI: the foundation of the smart grid
(b) Smart Homes and Networks
(c) Demand Response: a market primed to perform
(d) Network Reliability And Security (Self-Healing)
(e) Grid Optimization
(f) Integration of Renewable energy sources
(g) Asset Management and online Equipment Monitoring
(h) Participation in Energy Market.

4 India's Road Map for a Smart Grid Mission

Smart grid vision for India: transforms the Indian power sector into a secure, adaptive, sustainable, and digitally enabled ecosystem that provides reliable and quality energy for all with the active participation of stakeholders.

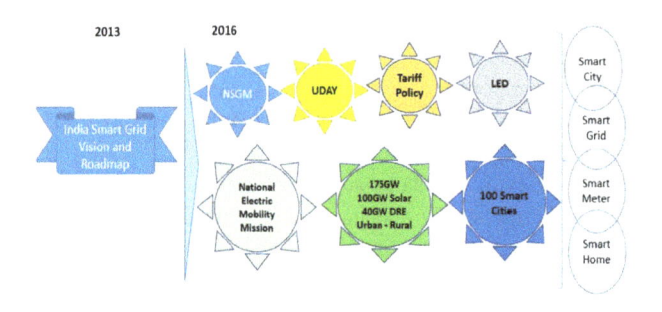

5 Challenges with Reintegration

  i. Reliability top priority
 ii. Variability and uncertainty in aggregate electric demand
iii. Unit commitment
 iv. Integration of intermittency nature of renewable generation
  v. Frequency control
 vi. Reactive power management
vii. Physical constraints (i.e., available transmission).

(a) *Intermittency and Variability*:

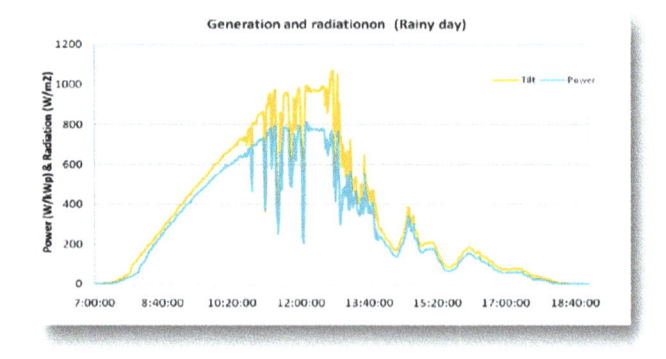

(b) *Non-predictability*

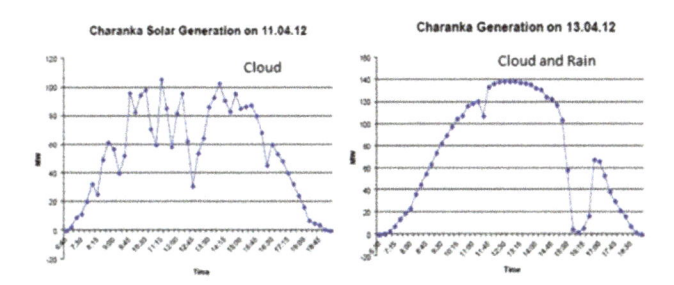

(c) *Location Specificity*

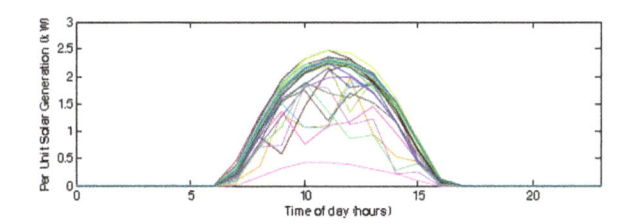

6 All India Wind Energy Generation

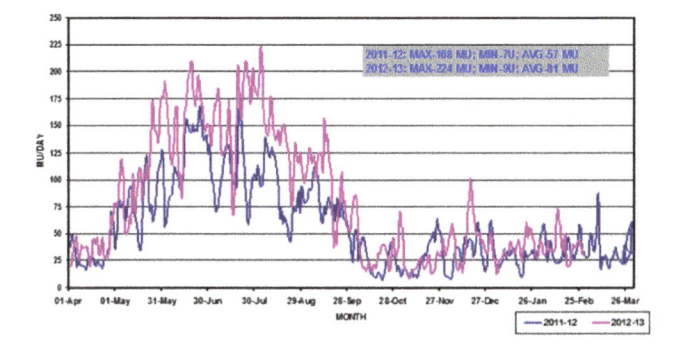

7 Integration of Renewables on the Load Curve

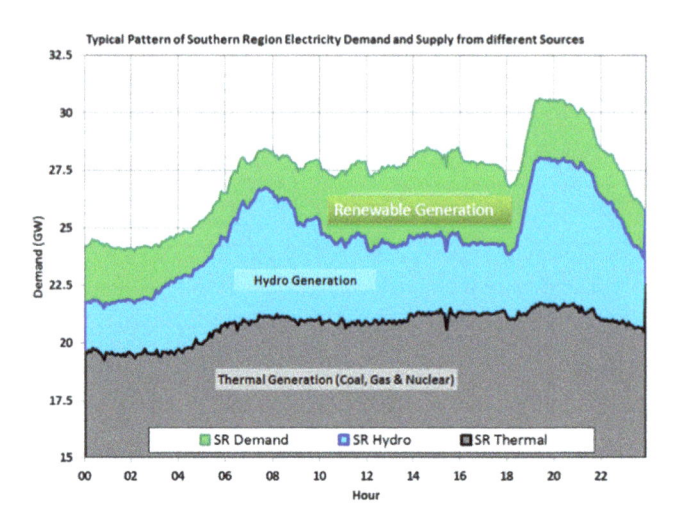

8 Smart Grid Solutions to Overcome Challenges Posed by Reintegration

(a) *Network reliability and Security*

- Synchrophasors and the network of PMUs are known as wide area measurements system (WAMS).
- Phasor measurement units (PMUs) enable us to understand the dynamic behavior of the power system.
- Modern technology for indication of stress on transmission and can be used to trigger corrective actions to maintain reliability.
- Smart grid makes use of technologies that improve fault detection and allow self-healing of the network without the human intervention.

(b) *Demand Response*

NET metering helps in bidirectional power flow and accountability. Consumer can contribute to demand response which can be a win–win situation for both utility and consumer.

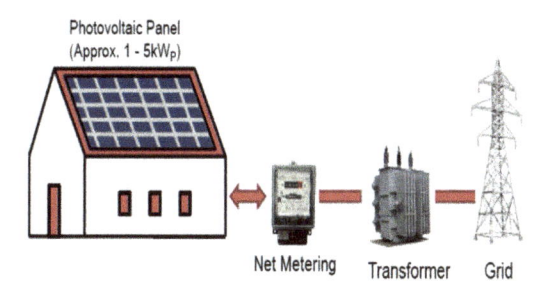

Peak shaving can be possible with the coordination between utilities and consumers, thereby resulting in postponement of heavy investments for power projects, brownouts/blackouts/reduced electricity bills, reduced purchase of costly power to meet peak demand, etc.

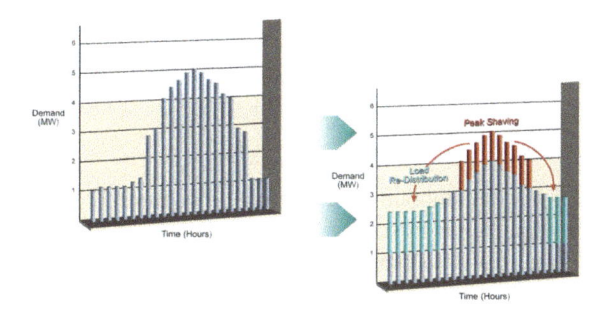

(c) *Prosumers*

- Customer engagement is not only limited to controlling their energy consumption
- But also to feed back to the grid
- Depending on the TOU rates and demand of the grid, prosumers may feed back to the grid by postponing their consumption to off peaks
- Benefit to prosumer is that he shall be credited for the amount of energy fed back to the grid.

(d) *Reintegration Through Microgrids*

- Main challenge with renewable is to maintain a stable and reliable power grid
- Any small-scale localized station with its own power resources, generation and loads and definable boundaries qualifies as a microgrid

- Renewable microgrids are individual rooftop solar/wind plants that can generate power and feed the main grid at times of excess generation/peak demand
- Smart grid provides data and automation needed to enhance renewable energy to put energy on the grid and optimize the usage.

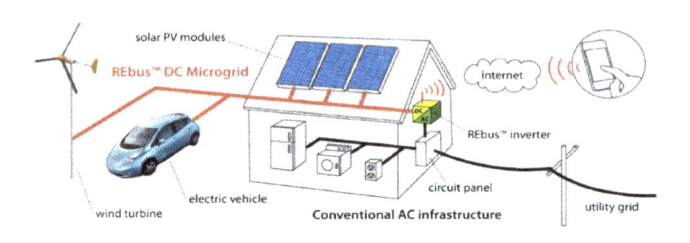

(e) *Smart Inverters with advanced power controls*

These and other power controls can reduce the need for significant grid transmission and distribution upgrades, thus reducing costs that may otherwise be levied on RE projects.

(f) *Integrated Storage*

- Storage can help to smooth short-term variations in RE output
- to manage mismatches in supply and demand
- pumped hydro, battery storage, flywheel, etc.

(g) *Behind the meter storage*

- Customer storage solutions can help absorb excess PV generation, reducing the need for distribution upgrades.

(h) *Advanced Energy Management systems*

- Provide real-time, high-resolution visibility and control of power systems, which can allow grid operators to defer more costly capital expenditures.

(i) *Better Forecasting*

- System-level forecasting can help system operators operate their grids more flexible, allowing more production to be accepted.

(j) *Dynamic Line rating*

- Real-time information about transmission line capacity
- Thus, allowing grid operators to extract more values from existing lines, and reducing the need for costly upgrades.

(k) *Real-time system awareness and management*

- Instrumentation and control equipment across transmission and distribution networks allow system operators to have
- real-time awareness of system conditions and
- ability to actively manage grid behavior.

9 Electric Vehicles: Need and Levers

10 Need for Electric Mobility

(a) Faster depletion of fossil fuels,
(b) Rapid increase in energy costs,
(c) Impact of transportation on the environment and concerns over climate change,
(d) India's primary energy consumption is expected to increase by 70% by 2022. Hence, gap between domestic crude oil production and consumption is widening,
(e) Transportation sector accounts for one-third of the total crude oil consumption. Road transportation accounts for 80% of this consumption.

11 Key Levers for Promoting Electric Mobility

(a) Demand incentives
(b) Supply-side incentives to spur manufacturing
(c) Power and charging infrastructure
(d) R&D initiatives
(e) Imposition of stringent fuel efficiency norms.

12 NEMMP 2020 Mission Overview

Electric Vehicles – Changing Landscape
National Hybrid/Electric Mobility Mission proposals

- **Develop a mission plan** for promoting a range of electric mobility solutions to enhance national fuel security, provide affordable and eco-friendly transportation
- **Mandate higher** fuel efficiency norms with penalties for non-compliance
- **Mandate electric** vehicles in government fleets, public transportation
- **Incentivise sales** of electric vehicles though cash subsidies
- **Provide OEMs** and suppliers with accelerated depreciation and tax holidays
- **Phase out** low import duties on components over 5 years to encourage localisation
- **Fund R&D programs** along with manufacturers to develop low-cost solutions
- **Roll out** pilots to support hybrid and electric vehicles
- **Make modest** investments to build public charging infrastructure to support electric vehicles

- Emerge as a leader in xEV for two- and four-wheeler market in the world by 2020.
- Total xEV sales of 6–7 million units and contributing toward national fuel security.
- By 2020, India may be the third largest vehicle market in the world.
- NEMMP 2020 is the basis for guiding all the future initiatives, schemes, policies, and other interventions of the government for electric mobility.

As per the policy, xEV Potential demand by 2020 is as below:

Vehicle Segment	Potential for xEVS (Million Units)
Battery EV – 2 wheeler	3.5 – 5
Hybrid EV – 4 Wheeler, Bus, Light commercial vehicle (LCV)	1.3 – 1.4
Other Battery EV (3 Wheelers, 4 Wheelers, Bus LCV)	0.2 – 0.4
Total	6 - 7

However, there are many barriers to achieve this mission. Some of the key barriers are listed below:

- Consumer acceptability
- Technology development
- Manufacturing investments
- Lack of xEV related infrastructure.
- Fast charging stations
- Range limitation (max distance driven per charge)
- Govt continued support for kick starting xEV, R&D
- Development of battery swap
- Viable business model
- Standardization of technical characteristics like battery characteristics, motor specifications etc

Efforts need to address electric vehicle-related infrastructure related issues, viz.

(a) Augmentation of existing power generation and transmission
(b) Setting up charging infrastructure for xEVs
(c) Particularly for buses—depots
(d) Other vehicle segments—apartments, malls, parking garages, workplaces, public buildings, etc.
(e) Charging stations shall be developed as a commercially viable business opportunity that attracts private investment
(f) Mandate charging infrastructure in public buildings (amend building laws)
(g) Introduce standards for charging equipment
(h) Allowing private retailing of power
(i) Uninterrupted electricity for xEVs recharge
(j) Target cities, fleet operators, public transportation companies, operators of feeder buses to metro systems, etc.

India's charging infrastructure at a glance:

	High Gas / HEV – 4W	High Gas / HEV/BEV – 4W	High Gas / HEV – Buses	High Gas / HEV/BEV – Buses
Extra generation required (MW)	150	225	2	4
Charging infrastructure (Crs)	750	1000	10	20
Charging terminals in 000	175	227	310	500

13 Business Opportunities from Electric Vehicles

(a) xEV charging infrastructure can be a self-sustaining viable business opportunity if adequately facilitated by the government.
(b) Most of the investments in charging infrastructure can be made by the private sector.
(c) Modification in electricity act is required for allowing charging stations to sell power. This will necessitate enabling legislation, modification to the relevant laws, and also perhaps legal mandates for private charging infrastructure.
(d) Government needs to ensure standardization of batteries and recharging-related components as this is an essential ingredient for successful rollout of recharging infrastructure.
(e) Public xEV charging infrastructure needs greater participation of power utilities along with battery suppliers and vehicle manufacturers.
(f) Introduction of standards for vehicle to grid interface, components, batteries, and other charging infrastructure is much needed to ensure minimum quality and safety standards and also to achieve economies of scale.
(g) Testing infrastructure for testing and certifying is also needed.
(h) Carrying out studies for impact assessment of xEV charging on the microgrids and possibility of using xEV to power/stabilize the grid and act as an energy storage medium in remote areas powered by renewable energy.

14 Conclusion

(a) Smart grid technology is new and evolving. Worldwide this technology is being tried and tested as pilot projects of various scales.
(b) Smart grid technologies enhance flexible operation of the grid.
(c) Improve utilization of existing infrastructure.
(d) Peak load management—demand side management for optimal utilization of assets.
(e) Two-way interaction between utility and a consumer.
(f) Demand response delaying huge investments.
(g) Improve grid stability by integrating EV with grids or use as microgrids.

References

1. Smart Grid Vision and Roadmap for India by Ministry of Power—12th Aug 2013
2. Smart Grids in Distribution Networks—Roadmap development and implementation by International Energy Agency (IEA)

3. Demand Side Management in India: Technology Assessment by SHAKTI Sustainable energy foundation
4. Radha Swaminathan, Smart Grid and Utility Transformation—A look at transitioning electric utilities into the 21st century
5. The Role of Smart Grids in Integrating Renewable Energy, ISGAN Synthesis Report
6. National Electric Mobility Mission Plan 2020, Department of Heavy Industry, GOI
7. Recommendations for Electric Vehicle Policy and Charging Infrastructure for Incorporating in the NEMMP Framework and Policies, ISGF (India Smart Grid Forum) and SUG (Smart Utilities Group)
8. Draft National Policy on RE based Mini/Micro grids, Ministry of New and Renewable energy (MNRE), 30th May 2012

Author Biography

Goutham Chakravarthy Yelmanchli Goutham Chakravarthy Yelmanchli is currently working as Lead Customer Acquisition for Tata Power Mumbai Distribution business.

Educational background: He is postgraduate with Master's in Business Management from NTPC School of Business with Gold Medal and Master of Technology in Power Systems specialization from College of Engineering, Pune.

Professional background: He has a professional experience of 8+ years in Design and Engineering profile of a distribution utility in Tata Power Company Limited, Mumbai.

Profile includes electrical and layout designing of substations, technical specifications, material management, capex planning, regulatory DPR preparations, etc.

He was a core team member representing distribution in SAP —Saarthi implementation project taken up by Tata Power. Some of the technical participation in his profile include Tata Innovista, ITMA, design, and retro-filling of ester-filled transformers.

Solar Agricultural Pump Without Electric Motor

G. V. Sukumara and Vijay L. Sonavane

Abstract Due to immense irrigation, the water table is depleting. In many parts of the country, the water table has gone down below 800–1000 ft. Multistage and booster pumps are used to pump the water. The conventional variable voltage and frequency submersible motor needs to be designed to pump water from such areas, and they also need capability to withstand the large voltage variation due to the voltage drop in the solar panel. Electrical wiring has to run from the panel outside to the motor inside the well. All these variables result in low efficiency as well as higher capacity solar panel. In addition to this, fine sand, slush, etc., that come along with the water damage the pump. It is a herculean task for a farmer to repair the pump-motor set. Though the government subsidizes the installation of solar pumps, the repair cost has to be borne by the farmer. One side, there is erratic monsoon, and on the other side, cost of maintenance and availability of qualified personals to repair have put the farmers in distress. Today, about 42.3% connected load to the grid is of that of the agricultural pumps. The probability of all pumps not working simultaneously has helped to manage the power demand. To overcome all the deficiency mentioned above, a totally a new concept of pumping the water from deep well as low as 1200 ft without electrical motor with less than 25% of solar panel capacity is discussed in this paper. Once the concept is implemented, the grid can be relieved from the agricultural load, leading to surplus power in the country for the next 20 years.

Mr. Sukumara, inventor of this innovative concept and the author of this paper, has applied for the patent, and the same is pending with permanent number 201641041397 dated 03/12/2016. Mrs. Vani Sukumar is the assignee of the patent. Inventor and author of pump Mr. G V Sukumara firmly believes that his intellectual property rights will be well protected in our country as well as globally.

G. V. Sukumara (✉)
Prapati, No 36, 2nd Main Road, Kannappa Nagar, Thiruvanmiyur
Chennai 600041, India
e-mail: project.prapati@gmail.com

V. L. Sonavane
MSEDCL, Mumbai, India
e-mail: vlsonavane@gmail.com

© Springer Nature Singapore Pte Ltd. 2018
R. K. Pillai et al. (eds.), *ISGW 2017: Compendium of Technical Papers*, Lecture Notes in Electrical Engineering 487, https://doi.org/10.1007/978-981-10-8249-8_22

1 Introduction

Following are the extracts from the article published by P. S. Vijay Shankar, India's Groundwater Challenge "CENTER FOR THE ADVANCED STUDY OF INDIA" on February 28, 2011, which briefs about groundwater situation in India.

The most dramatic change in the groundwater scenario in India is that the share of bore wells in total irrigated areas went up from a mere 1% during 1960–61 to 40% during 2006–07. The estimated number of wells and bore wells in India is now around 27 million, with bore wells accounting for more than 50%. On average, there were 27 bore wells per square kilometer of sown area in Punjab, 22 in Uttar Pradesh, and 14 in Haryana. Interestingly, small and marginal farmers (with landholding sizes of less than two hectares) accounted for over two-thirds of the holders that own bore wells [1]. With 27 million wells and bore wells if electrified pump used at the average rate of 5 kW per pump, the total connected load is about 131 million kW, i.e., 131 GW. The installed capacity of power generation as on December 30, 2016, is 310.01 GW [4].

Power distribution for agricultural pumps is either free or subsidized. To meet the peak power demand, agricultural feeders have been separated. Agricultural pumps get power for 8–10 h in a day.

2 Impact of Subsidized or Free Power Supply on Farmers

Following are the extracts from an article "Do Farmers Need Free Electricity? Implications for Groundwater Use in South India" published by Shri Elumalai Kannan, Institute for Social and Economic Change, Agricultural Development in Journal of Social and Economic Development, July–December 2013.

Although electricity is supplied free of cost to the agricultural sector, farmers were not happy with the quality of electricity supplied in terms of hours of supply per day and faced frequent voltage fluctuations, which resulted in high cost of maintenance of pump sets. sample farmers in both the states of Tamil Nadu and Karnataka power availability is about 4–5 h a day with heavy voltage fluctuations. Because of heavy voltage fluctuations in the study area, farmers incurred average repair cost of Rs. 6000 per year. Frequent power cuts and interrupted supply were the other common problems reported by the sample farmers. These problems often result in pump set burnouts. Pump set burnouts occur not only due to the poor quality of electricity, but also due to inefficient cheap pump sets (locally rewound) available in the market.

Since electricity for agricultural pumps has been highly subsidized or free in most of states, there was no incentive for manufacturers to produce good quality and environment-friendly pump sets (Sant and Dixit 1996) [2].

3 Solar Water Pumping for Irrigation Potential and Barriers in Bihar, India

Market potential for solar water pumps in India
The Centre for Study of Science, Technology and Policy (C-STEP)
Estimates 9 million diesel water pumping sets in use in India. If 50% of these diesel pumps were replaced with solar PV pump sets, diesel consumption could be reduced to the tune of about 225 billion liters/year.

Hamburgisches Welt Wirtschaftsinstitut (HWWI) estimates the potential to be 70 million solar PV pumps by 2020 (14 million in Uttar Pradesh & 11 million in Bihar).

KPMG estimates approximately 16,200 MW as the total potential in the agricultural sector by 2017–22.

	Barriers	Potential Solutions
Market Related Barriers	High Upfront Cost	Smart Subsidies/Innovative Finance
	Lack of Finance Mechanisms	Innovative Consumer/Business Finance Mechanisms
	Low Awareness Among Consumers & other Relevant Stakeholders	Awareness Campaigns
	Lack of Maintenance and Support	Localised Service Infrastructure
	Lack of Market Intelligence and Information	Provision of Adequate Resources/Market Data
	Danger of Theft	Portable/Community Owned Systems, Insurance
Regulatory Issues	Restricted Financial Engineering	Innovative Policies and Financial Engineering
	Maze of Political Departments	"Single-Window" Approach
	Lack of Market-Oriented Policies	Policies providing a Level Playing Field with diesel pumps
	Concealed Tenancy and Small Landholdings	Tenancy Reform, Leasing Mechanisms & Group Investments
Technology Related Barriers	Lack of Standardisation & Quality Assurance	Standardised Products that Cater to Local Needs
	Lack of local Manufacturers	Promotion of Local Manufacturing

3.1 Technology-Related Barriers

1. *Solar PV array*: The solar PV array is a set of photovoltaic modules connected in series and possibly strings of modules connected in parallel.
2. *Controller*: The controller is an electronic device which matches the PV power to the motor and regulates the operation of the pump according to the input from the solar PV array.
3. *Pump set*: Pump sets generally comprise of the motor, which drives the operation and the actual pump which moves the water under pressure.

3.2 Our Invention for Cost-Effective Solution

Our solution is to address the above problems faced by the farmers, government, distribution companies, and other stakeholders.

It is sincerely felt that in order to achieve end-to-end solution, our country needs manufacturers of solar water pumps who work with drip irrigation experts and water table experts to give commitment to supply a minimum amount of water annually for a minimum period of 5 years.

Keeping the need of rural population and agricultural farmers in our country (and globe), Prapati, a knowledge-based Research and Development Company in Chennai, has developed "Solar-Powered Water Pumps without controller, without electric motor," without wires running inside the bore well, and with many more advantages called "Shashiratna" based on a new concept of "*Direct lifting pump with positive displacement of water.*"

3.3 Direct Action Hand Pumps

Direct action hand pumps have a pumping rod that is moved up and down, directly by the user, discharging the water. Direct action hand pumps are easy to install and maintain, but are limited to the maximum column of water a person can physically lift (up to 15 m).

3.4 Deep Well Hand Pumps

Deep well hand pumps are used for high lifts of more than 15 m. The weight of the column of water is too high to be lifted directly, and some form of mechanical advantage system such as a lever or flywheel is used.

High-lift pumps need to be stronger and sturdier to cope with the extra stresses. A deep well hand pump theoretically has no limit, to which it can extract water. In practice, the depth is limited by the physical power a human being can exert in lifting the column of water, which is around 80 m.

The said pump is "direct water lifting pump with positive displacement of water." There are two tubes: the inner tube and the outer tube.

Both inner tube and outer tube travel up and down. Both the tubes are connected together through a chain drive. Both the tubes travel in opposite direction. If inner tube travels upward, at the same time outer tube travels downward, and vice versa. One-way valve which normally allows water to flow through it in only one direction is connected at the bottom end of the inner tube as well as at the bottom end of outer tube. The inner tube with one-way valve travels from zero mm at the bottom to 300 mm from the bottom to top. Outer tube travels 300 mm from the bottom to 600 mm from the bottom to top (stops at 300 mm from the bottom). When the inner tube is at the bottom that is zero mm from the bottom, the outer tube will be at the top, that is, 600 mm from the bottom.

When the outer tube is traveling upward, the one-way valve closes and lifts the water column upward and discharges the water. At the same time, the inner tube's one-way valve opens and allows the tube to slide down comfortably. At the end of the travel, the 300 mm column of water is discharged and the inner tube reaches the bottom. Now, the inner tube starts moving upward, the one-way valve closes and starts lifting the water column, and outer tube's one-way valve opens and starts sliding comfortably. The inner tube moves the water column by 300 mm, the space cleared by discharging the water in the previous cycle will be filled with water in this cycle, and this is positive displacement.

The inner tube stops at 300 mm from the bottom, and outer tube also stops at 300 mm from the bottom, that is, the meeting point of both the tubes. The outer tube discharges 300 mm column of water from bore well at every alternate cycle (will be improved to every cycle after some trials). Inner tube displaces 300 mm column of water and fills the space that is created by the outer tube by discharging the water column at every alternate cycle.

Deep well–direct lifting pump theoretically has no limit, to which it can extract water. In practice, the depth is limited by the capacity to lift the water column. In this design, there is a counterweight whose weight is equal to the weight of water column and the weight of water which is being discharged for every alternate cycle. The weight is mounted in concentric position, and the weight gets transferred from inner tube to outer tube and outer tube to inner tube at every alternate cycle. This counterweight is moved up by 300 mm at every cycle with a speed of 2 m per second. This movement takes place in less than 150 ms.

Magnetic levitation (Maglev) is a *Magnetic force* used to counteract the effects of the *acceleration* of gravity. To develop magnetic force, copper or aluminum wound coil and core made of magnetic material such as cold-rolled grain-oriented (CRGO) silicon steel material is used.

The core is inserted inside the coil. The coil is energized for a short duration of about 50 ms, by passing the large current through the coil. The coil develops a tremendous force for a short time. The core is connected to counterweight. The counterweight moves up without any friction. This force overcomes the inertia and lifts the counterweight with a speed of 2 m/s.

Once it is thrown up, it moves due to its own inertia. The counterweight will get locked, when upward force is just equal to gravitational force and the weight is about to move down, due to gravitational force. This counterweight (when it sits on the inner tube or outer tube) moves the tube downward.

The tubes are connected through a chain and made to travel in opposite direction. When inner or outer tube travels downward, the other tube starts moving upward. The water column on the other tube starts moving up, and the weight of water gets transferred to the tube moving downward through the chain. Water gets discharged at every alternate cycle, and the water gets positive displacement and fills the space cleared by discharge at every other alternate cycle.

Solar panel is a high impedance source. The moment the current is drawn, the voltage of the panel drops. The capacitor bank gets charged by the solar panel. As the capacitor gets charged, the current drawn by the capacitor bank drops. As the current drops, the voltage of the solar panel increases and the capacitor charges to a higher voltage.

Both capacitor bank and solar panel work so well that peak charging takes place automatically and maximum power point tracking (MPPT) is not required. A small controller will give 50-ms pulse to the power electronic switch such as MOSFET, IGBT, and ISBT, which will energize the coil for a short duration of 50 ms (no experts are needed to maintain the panels). The number of pulses generated to power the coil depends upon the capacity of solar panel to charge the capacitors. If the solar panel power is low, then the number of pulses per minute will be less. Hence, the pump will work even if the solar output is low.

3.5 For Example

A 1250 W solar panel at 200 V will charge capacitor banks with 6.25 A for 2.5 s, i.e., 15.625 ampere-second at the end of 2.5 s.

The same 15.625 ampere-second is discharged through the coil in 50 ms, i.e., 312.5 A in 50 ms. For the coil of 15,000 turns, 4.687 million ampere-turns for 50 ms develop enormous force that can throw even 3000 kg counterweight up for 50 ms, and due to its own inertia, the counterweight will move more than 300 mm in less than 150 ms.

The frequency of energizing the coil will depend on the capacity of solar panel to charge the capacitor. The pump will work even with a very small power output of solar panels. The pump can work from during daytime 8 AM to 6 PM, i.e., for 10 h, when the solar PV panel is active (Fig. 1).

The pump consists of the following:
(1) solar panel, (2) control box, (3) coil and core combination for magnetic lifting, (4) deadweight, (5) locking mechanism 1 for inner tube, (6) locking

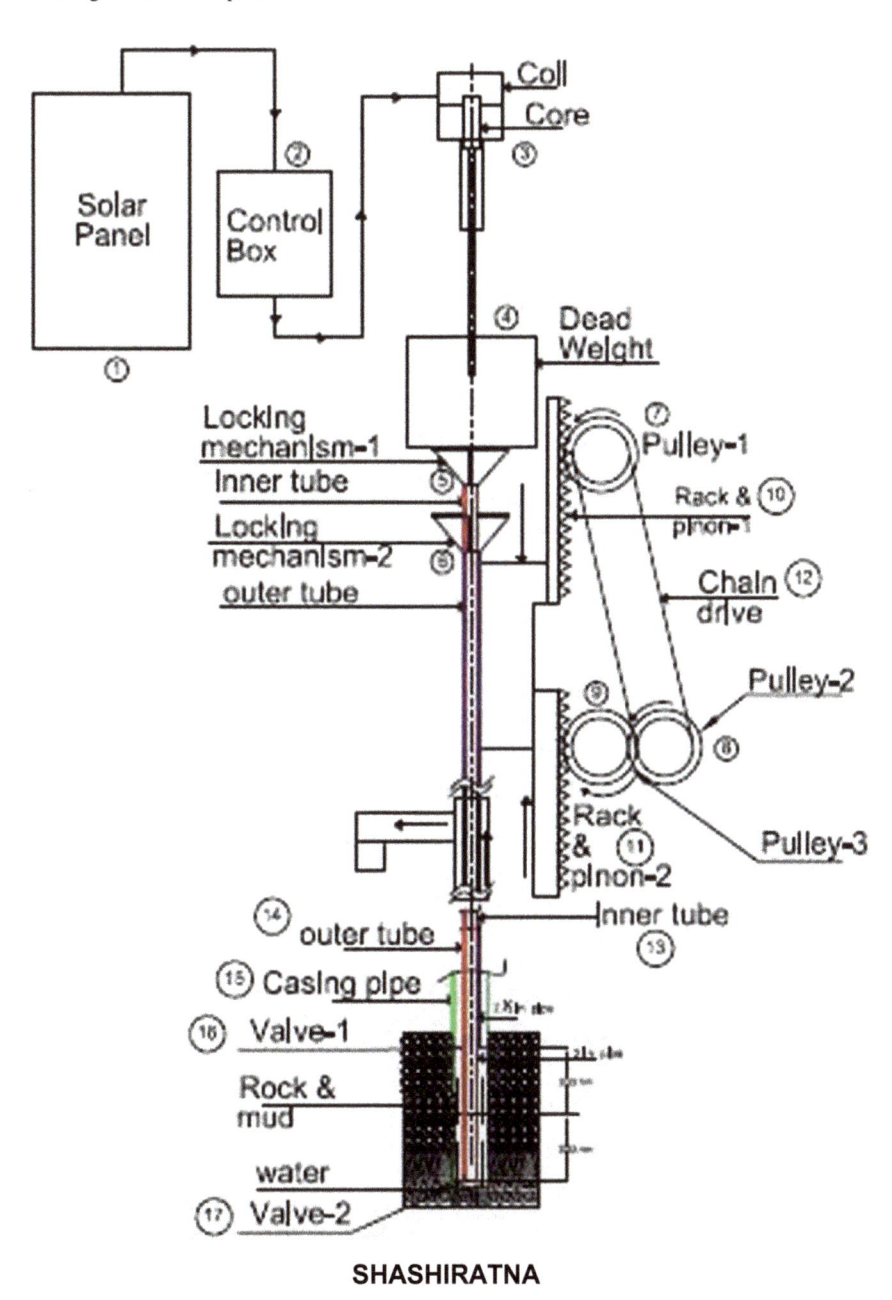

Fig. 1 Operation of pump

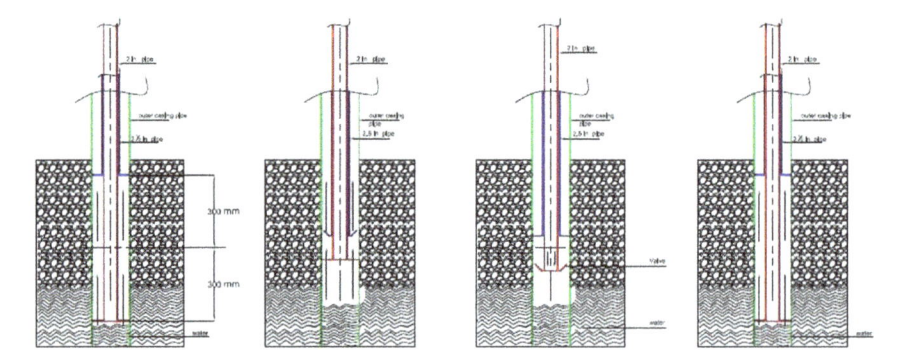

Fig. 2 Function of pump

mechanism 2 for outer tube, (7) pulley 1, (8) pulley 2, (9) pulley 3, (10) rack and pinion 1, (11) rack and pinion 2, (12) chain drive, (13) inner tube, (14) outer tube, (15) casing pipe, (16) valve 1, and (17) valve 2.

Both inner tube (13) and outer tube (14) travel up and down. Inner tube is connected to outer tube through pulley 1 (7), pulley 2(8), and pulley 3 (9), rack and pinion 1(10), rack and pinion 2 (11), and a chain drive (12). One-way valve 1 (16) is connected to outer tube. One-way valve 2 (17) is connected to inner tube. Deadweight (4) is connected to core and coil arrangement (3) to develop magnetic force to push deadweight upward. Locking mechanism 1 (5) locks the deadweight to inner tube. Locking mechanism 2 (6) locks the deadweight to outer tube. Casing pipe (15) is the pipe through which the water flows. Control box (2) generates suitable pulse to develop magnetic force. Solar panel (1) powers the controller (Fig. 2).

3.6 Operation of the "Direct Lifting Pump with Positive Displacement of Water"

At start, inner tube will be at bottom and outer tube will be at 600 mm above the bottom of the well.

Casing pipe from bottom of the well to the surface will be filled with water. Valve 2 connected to the inner tube will be closed. The deadweight is transferred to outer tube. Due to weight, outer tube starts coming down. The valve-1 connected to outer tube opens and facilitates the outer tube to slide comfortably. The pulleys 1, 2, 3, rack and pinions 1, 2 and chain drive will make inner tube to travel upward. The Valve 2 connected to inner tube closes, and the entire water column starts moving up. Thereby, the load gets transferred to outer tube which is pushed by counterweight.

The inner tube travels upward by 300 mm, and outer tube travels down by 300 mm. At this time, the coil will be energized and the magnetic force throws the deadweight up, without any friction to the top. Now, the deadweight gets locked to inner tube. Due to deadweight, the inner tube starts traveling down. The pulleys 1, 2,3, rack and pinions 1,2 and chain drive will make outer tube traveling upward. The one-way valve connected to outer tube will close and start lifting the water column upward.

Since there is no load on inner tube and the valve 2 connected to inner tube opens, the tube starts gliding down comfortably. When the inner tube reaches the bottom of the casing, it is already filled with water and ready to move. Again, the deadweight moves up and the outer tube discharges the water out. In the first stroke, the water column moves up by 300 mm and this is positive displacement of water.

In the second stroke, the water will discharge to the tank. The amount of water depends on the movement of deadweight which depends on the capacity of the solar panel charging the capacitor.

Core and coil arrangement with control box will make deadweight to move up by 300 mm at any determined time. Solar panel will power the coil at the desired interval.

Assume that the diameter of pipe is equal to 125 mm, and the outer tube diameter is 50 mm.

The net area of cross section will be 3.14 × 12.5 × 12.5/4 − 3.14 × 5 × 5/4 = 122.6 − 19.6 = 103 cm^2, and the volume of water per meter will be 10.3 L.

If depth of well is 810 ft (810/3.28 = 246 m), the volume of water and weight of water will be 2470 L and 2470 kg.

The volume of water which discharges every 5 s will be 10.3 × 300/1000 = 3.09 L, i.e., 3.09 × 12 × 60 = 2224.8 = 2225 L/h. And from 8 AM to 6 PM (10 h) during daytime, total volume of water will be 2225 × 10 = 22,250 L per day (if we take the suction of water during movement of the tubes, then there is a scope for doubling the discharge of water without any change and this has to be confirmed only after proper trials). 1.25 kW panel gives an output of 1250/200 = 6.25 A at 200 V. Power available at every 2.5 s will be 6.25 × 2.5 × 1000/50 = 312.5 A for 50 ms.

This will generate 312.5 × 15,000 = 4.687 million ampere-turns, which will develop sufficient force to move 2500-kg deadweight for 300 mm. The impact of weight while lifting the deadweight and while transferring from one tube to the other will not be on the tubes.

One mobile solar panel can power two-to-three pumps in the same area depending on water recharge, so that the final cost can be brought down substantially. Particularly for the farmers who has two-to-three hectares of land and have dug two-to-three bore wells, the cost of installation comes down drastically. The dream to delink the grid from agricultural load can be easily achieved by this system. 6-HP 45-stage submersible pump with 6 kW solar panel is needed for delivering 22,250/8 = 2781 L/h for 8 h at a head of 810 ft.

3.7 Conclusions and Challenges Ahead

This is an innovative approach to solve the problems faced by the farmers having electrified agricultural pumps in their farms. At present, they are struggling with very low-quality power supply with frequent interruptions. The grid connectivity for agricultural consumers is also costly affair. Hence, the innovative idea of "*Direct lifting pump with positive displacement of water*," a self-sufficient system, through which farmers will get 10 h of water supply for their farms, without putting burden on the grid system of Discom, has been developed by M/s PRAPATI.

At present, TEAM PRAPATI is working on resolving the following challenges:

(1) Design and production of core, coil arrangement to move 2500 kg of deadweight for 300 mm at a speed of 2 m/s.
(2) Design and production of prototype of inner tube, outer tube, pulleys, chain arrangement, one-way valve, etc., with a load of 2500 kg.
(3) Locking arrangement to lock the weight at the right moment when the force acquired due to mass and acceleration of the deadweight moving up is equal to the force due to acceleration of gravity.
(4) Smooth transfer of deadweight from inner tube to outer tube and vice versa.

3.8 Future Plans

Once the product is designed, the technology can be transferred to various participating companies at district level to supply, erect, and commission the system.

These companies can work in association with Drip Irrigation Department and Water Table Managers (Irrigation Department) and develop a plan to give commitment to supply a minimum quantity of water per annum for a period of 5 years to the farmers.

TEAM PRAPATI also have an idea to store the excess solar energy in pressurized vessels, so that using air turbines, the electricity generated can be used to power other needs of farmers, at their residences for domestic lighting, fans, home TV, computers, small refrigerators, etc.

3.9 Need Support from Various Organizations

PRAPATI requests all proactive organizations (Government in Power/Agricultural Ministry, CEA CERC and SERCs/public: electricity utilities/banks, etc.), institutions (IIT/engineering colleges/BIS/NABL laboratories), and forums (India Smart Grid Forum, FICCI) to recognize the innovative effort taken by our technology team and requests all these organizations to extend proactive support in terms of technology development, finance, etc., for setting up pilot projects all across the country. Our final aim is to extend farmers of our country, the required quantity of water for irrigation and drinking as well as good quality assured power supply and to relieve the distribution utilities from supplying power to ever-increasing demand of agricultural sector. We are sure that our innovative scheme is going to bring real "GREEN REVOLUTION" in the country.

This product is invented to support the farmers and the government of our MOTHER INDIA AND GLOBE, and we need a strong support from you all.

Acknowledgements At the time of this publication PRAPATI has made substantial progress on this project. They have been successful in proving the concept by making a working model. Commercial model of 600 feet pump will be available by end of June 2018.

References

1. https://casi.sas.upenn.edu/iit/shankar
2. http://www.academia.edu/7710263/Do_Farmers_Need_Free_Electricity_Implications_for_Groundwater_Use_in_South_India
3. http://www.igen-access.in/download/publication/1471333847.pdf
4. https://en.wikipedia.org/wiki/Electricity_sector_in_India

Applicability of Error Limit in Forecasting and Scheduling of Wind and Solar Power in India

Abhik Kumar Das

Abstract Forecasting of power generation is an essential requirement for high penetration of variable renewable energy in existing grid system as the major purpose of forecasting is to reduce the uncertainty of renewable generation, so that its variability can be more precisely accommodated. This paper focuses on the statistical behaviour of error in solar and wind power forecasting considering Indian regulations and analyses the applicability of the error limit in calculating the energy accuracy of forecasting and the stability of the grid.

Keywords Forecast error · Variability · Penalty due to deviation

1 Introduction

One of the major challenges in renewable enabled smart grid is to ensure the demand–supply balance for electricity. Wind and solar energy are two major components of renewable energy generation in India [1–4], but the variability and unpredictability inherent to wind and solar create a threat to grid reliability due to balancing challenge in load and generation. The unscheduled fluctuations of wind and solar generation produce ramping events, and hence, the integration of significant wind and solar energy into existing supply system is a challenge for large-scale renewable energy penetration [5–11]. To accommodate the variability, the day-ahead and short-term renewable energy forecasting is needed to effectively integrate renewable energy to the smart grid and hence the forecasting and scheduling of wind and solar energy generation has become a widely pursued area of research at Indian context [12, 13].

The concept of forecasting and scheduling (F&S) of renewable energy generators and the commercial settlement was introduced in Indian context by CERC through Indian Electricity Grid Code (IEGC), 2010 [14] and the Renewable

A. K. Das (✉)
Del2infinity Energy Consulting, Kolkata, India
e-mail: abhik@del2infinity.xyz; contact@del2infinity.xyz

Regulatory Fund Mechanism [15]. Due to several implementation issues, the mechanism was never made operational. To formulate an implementable framework, CERC also issued the second amendment to regulation for deviation settlement mechanism and other related matters [16]. After CERC regulation, Forum of Regulators (FORs) [17] and other state regulators issued or drafted regulation related to the forecasting and scheduling of wind and solar power generation. Since the system operators in India have to do curtailment on variable renewable energy due to intermittency and variability of the wind and solar power generation, the forecasting takes an important role in creating a sustainable solution for maximum utilization of renewable energy.

The most usable and conventional strategies in F&S of wind and solar power generation are used to predict the weather parameters using Numerical Weather Prediction (NWP) models and convert the values of weather parameters into power generation using the turbine or PV models considering the CFD-based analysis in local regions. The recent development of deep learning algorithms in artificial neural network (ANN)-based methodologies has created a huge scope in forecasting the power generations. Considering the uncertainty in the initial value vector in NWP and learning vectors in deep neural network (DNN), a powerful perspective regarding forecasting methodology is to regard it fundamentally as a statistical rather than deterministic solutions as the stability of the grid needs not just the prediction of power generation but also the uncertainty associated with it. Thus, from a mathematical viewpoint, forecasting is best considered as the study of the temporal evolution of probability distributions associated with variables in the power generation. Forecasting with proper uncertainty analysis is required since the initial value vector and the neural architecture of DNN can never be precisely defined and there are observational limitations in different variables required to predict the power generation. Without proper forecasting models, considering nonlinear dynamics approaches, a small but appreciable fraction of error can create a large error due to temporal evolution.

In this paper, a simple functional relation is established considering the variability of power generation and error limit in forecasting for different regulations. The average penalty due to deviation is calculated using the exponential model of forecast error distribution. The remaining of the paper is organized as follows: Sect. 2 briefly describes the forecast error, and the functional relationship of variability of actual power generation is derived in Sect. 3. Section 4 shows the computational model of penalty due to deviations. The results and brief analysis are discussed in Sect. 5, and conclusion is presented in Sect. 6.

2 Forecast Error

The most usable mathematical techniques in defining the forecast error are mean absolute error (MAE) and root mean square error (RMSE). But considering the system stability, the error at each time block (1 Time Block = 15 min) has more practical consequences and the error at ith time block is defined as [15, 17]:

Table 1 Generalized structure of deviation charge

Error band	Deviation charge per kW-Hr (Rs)	
	PPA based [15]	Fixed (depends on regulations)
$e \leq m$	No-penalty	0
$m < e \leq m_1$	10% of PPA	INR 0.50
$m_1 < e \leq m_2$	20% of PPA	INR 1.00
$e > m_2$	30% of PPA	INR 1.50

$$e(i) = \frac{1}{Avc} |x_A(i) - x_S(i)| \quad (1)$$

where AvC is the available capacity; $x_A(i)$ and $x_s(i)$ are actual generation and scheduled generation, respectively. Considering the Indian regulation, the penalty due to deviation can be generalized as shown in Table 1 [13].

3 Variability in Power Generation

Variability of power represents the change of generation output due to unscheduled fluctuations of wind velocity or sun radiation patterns, while uncertainty describes the inability to predict in advance the changes in generation output. Large unscheduled changes in wind or solar power generation are called ramp events which hamper the penetration of variable power in the existing grid [13]. The variability can be quantified as a measure of dispersion in variable renewable power generation and be easily quantified using the concept of Lorenz curve [18]. Though there are different ways to quantify variability, the simplest way to measure variability is using the normalized squared deviation about mean of the actual generation for some time block n as follows:

$$\sigma_A^2 = \frac{1}{nAvc} \sum_{i=1}^{n} (x_A(i) - \overline{x_A})^2 \quad (2.A)$$

Similarly, the dispersion in the schedule generation for the same time span can be represented as:

$$\sigma_s^2 = \frac{1}{nAvc} \sum_{i=1}^{n} (x_s(i) - \overline{x_s})^2 \quad (2.B)$$

where $\overline{x_S}$ and $\overline{x_A}$ are the mean of the power generation in the time span of n. For a good forecast in which $\overline{x_S} \to \overline{x_A}$ and $\sigma_A \to \sigma_S$, we can show that (A.1).

Fig. 1 Relation between correlation coefficient of actual and schedule generation and the variability of the actual power generation which the forecast model effectively accommodates

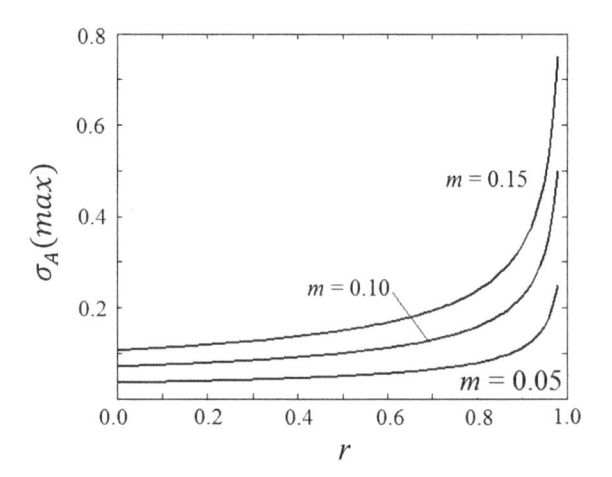

$$\sigma_A \leq \frac{m}{\sqrt{2(1-r)}} \tag{3}$$

where r is the correlation coefficient of the schedule and actual power generation. The relationship (3) shows how much variability a forecast model can effectively accommodate considering that the no-penalty is 100 m%, the value of which is defined by the regulation. As shown in Fig. 1, for some value of r, the maximum permissible variability of the forecast model can be calculated for different regulations, i.e. no-penalty error limit as 5% ($m = 0.05$), 10% ($m = 0.10$) and 15% ($m = 0.15$). The value of r depends on the choices of models and methodologies used in forecasting.

4 Penalty Due to Deviation

The penalty per available capacity can be represented as:

$$C = \int ec(e)h(e)\,de \tag{4}$$

where $c(e)$ is the penalty due to the error and $h(e)$ can be represented as the probability of error in the forecast models. Here, considering the Indian regulations, for simple statistical calculation we can assume that

$$\begin{aligned} c(e) &= k(e-m) \quad \text{if } e > m \\ &= 0, \quad \text{otherwise} \end{aligned} \tag{5}$$

Considering the phenomenological models of forecast error, the probability distribution can be viewed as an exponential distribution of parameter λ as follows:

$$h(e) = \lambda \exp(-\lambda e) \qquad (6)$$

Figure 2 shows the frequency plot of forecast error in solar and wind forecasting for different days. The frequency plot closely matches the exponential distribution defined in (6) for different values of λ.

Using mathematical manipulation, we can state that (A.2)

$$C \approx k\left[(m+\sigma_e)^2 + \sigma_e^2\right](1 - P(m)) \qquad (7)$$

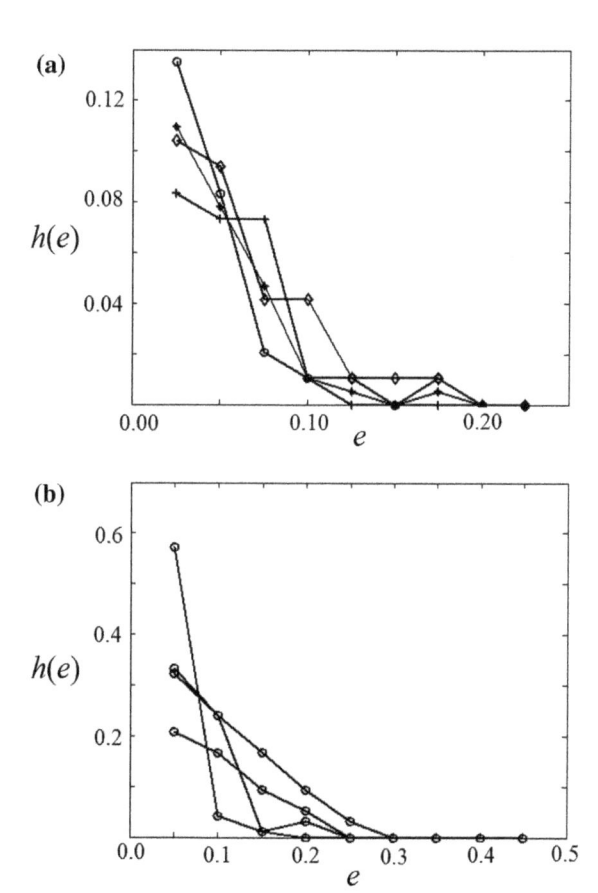

Fig. 2 Normalized frequency distribution of forecast error for different days of **a** solar forecasting and **b** wind forecasting

where $P(m)$ is the probability of error under no-penalty error band, i.e.

$$P(m) = \text{Prob}(e \leq m) = \int_0^m h(e) \mathrm{d}e \tag{8}$$

And σ_e is defined as the standard deviation of the error distribution. The average penalty due to deviation is always positive (Figs. 3a and 4b) and can be zero when $P(m) = 1$, i.e. prediction of power generation with 100% accuracy (As shown in Figs. 3a and 4b).

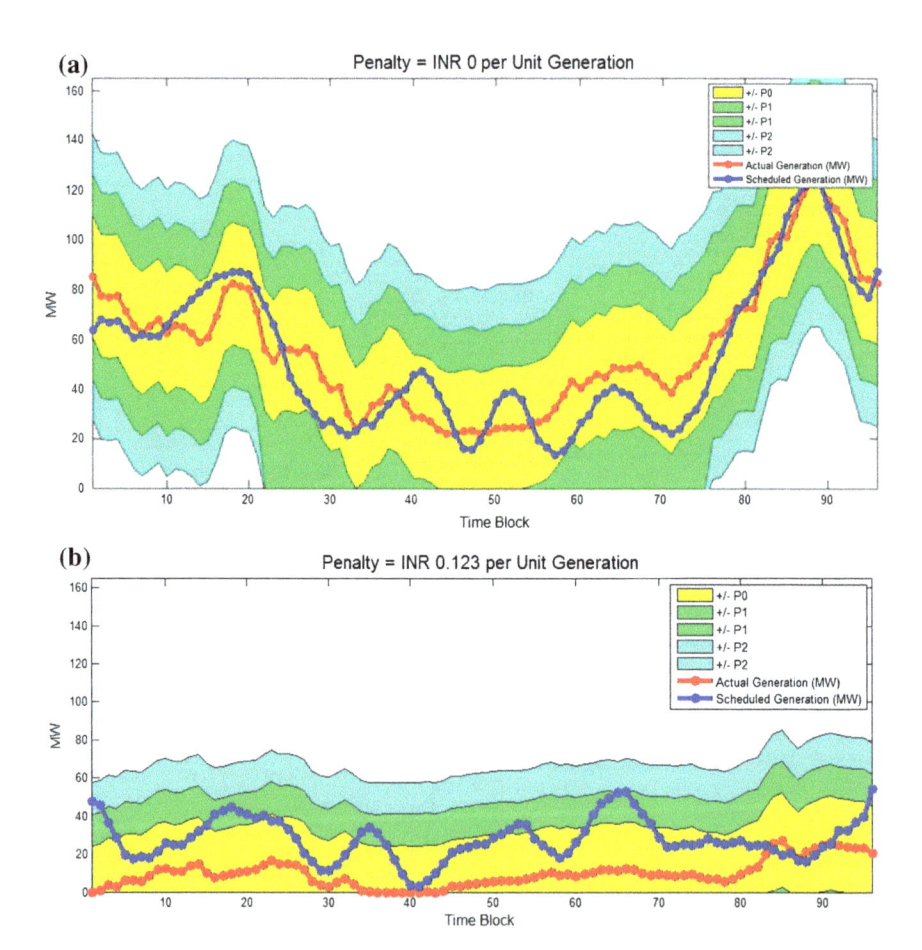

Fig. 3 Actual and forecast power generation of wind and its penalty due to deviation per actual generation

Applicability of Error Limit in Forecasting ...

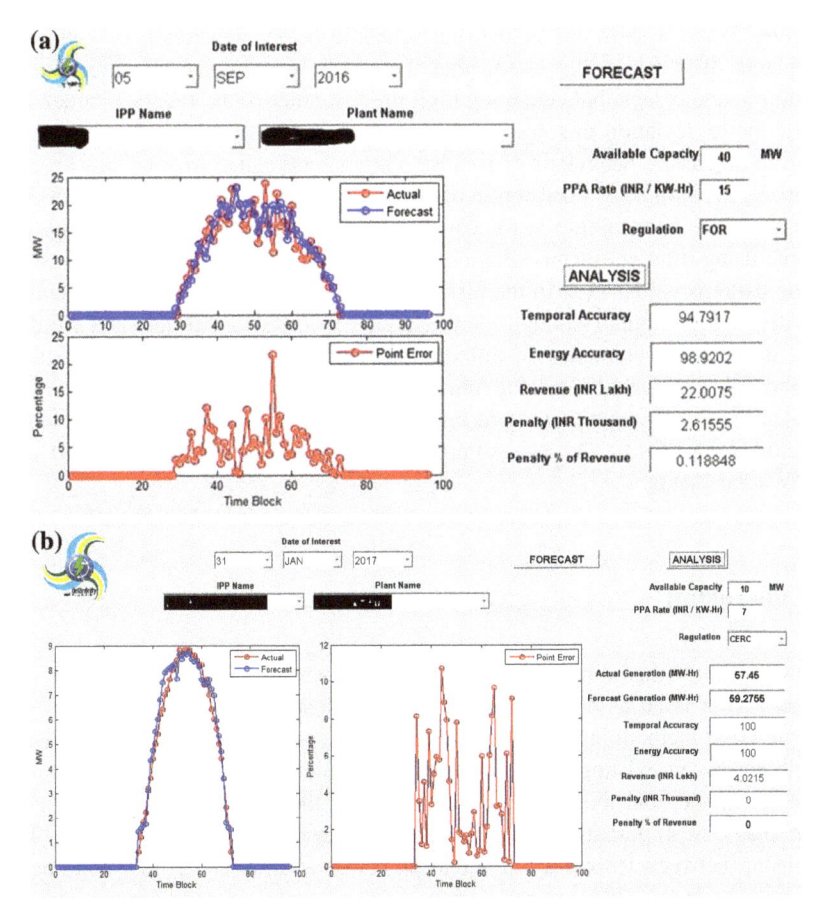

Fig. 4 Actual and forecast power generation of solar and its penalty due to deviation as percentage of revenue

5 Results and Analysis

Equation (3) represents an interesting relationship connecting physical parameter of the power generation (σ_A), regulatory parameter (m which is defined in Table 1 as no-penalty error limit) and accuracy parameter (r which is defined as correlation coefficient) of power generation forecast. This relationship shows the ability to forecast model to accommodate the maximum variability in the actual power generation considering different regulations (m) as the accuracy of the forecast model depends on the variability of the actual power generation.

As an example, Fig. 3 shows the daily forecast of a 160 MW plant in India; though there are variations in the power generation, the del2infinity's AI-based algorithm is useful to forecast the pattern of fluctuations.

Figures 3 and 4 show the actual and schedule power generation of wind and solar plants. It is interesting to see that though there are some penalties due to deviation in some days, but achieving high predictability of the forecast model, the penalty due to deviation in some days can become zero.

Due to stochastic behaviour of weather parameters related to solar power generation and the variability distribution of power, the temporal evolution of predicted variables creates uncertainties in forecasting. Though numerous forecast models are available using different methodologies, a good forecast is that which captures the genuine patterns which exist in the historical data, but do not replicate past events that will not occur again. Though there is a paucity of historical data and the data related to weather parameters of different solar plants in India, the del2infinity's AI-based solution can be used for forecasting and scheduling solution which can minimize the penalty due to present regulation and can work even when there is unscheduled variability. As an example, Fig. 4 shows the forecast of 40 and 10 MW solar plants.

6 Conclusion

Forecasting and scheduling (F&S) of variable renewable power generation like wind and solar is an essential requirement for a sustainable energy future as the proper forecast methodology reduces the uncertainty and helps to accommodate the variable energy in existing grid. The temporal evolution of predicted variables is non-deterministic, and the statistical prediction of the distribution of power generation plays an important role in forecasting with acceptable uncertainties. The relationship between variability of actual power generation and correlation coefficient in forecast for different regulations can be useful in forecasting, and the statistical formulation of average penalty due to deviation can be used in analysis of different forecast strategies.

Appendix

$$e(i) = \frac{1}{Avc}|x_A(i) - x_S(i)| = \frac{1}{Avc}|(x_A(i) - \overline{x_A}) - (x_S(i) - \overline{x_S}) + (\overline{x_A} - \overline{x_S})| \quad (A.1)$$

For a good forecasting system, we can consider $\overline{x_S} \to \overline{x_A}$ and $\sigma_A \to \sigma_S$. Hence, using the no-penalty band in Table 1, for some n we can state that

$$m^2 \geq \frac{1}{n}\sum_{i=1}^{n} e^2(i) = \left(\frac{1}{AvC}\right)^2 \frac{1}{n}\sum_{i=1}^{n}((x_A(i)-\overline{x_A})-(x_S(i)-\overline{x_S}))^2$$

$$= \left(\frac{1}{AvC}\right)^2 \frac{1}{n}\sum_{i=1}^{n}\left[(x_A(i)-\overline{x_A})^2+(x_S(i)-\overline{x_S})^2-2(x_A(i)-\overline{x_A})(x_S(i)-\overline{x_S})\right]$$

(A.2)

$= \sigma_A^2 + \sigma_S^2 - 2r\sigma_A\sigma_S = 2(1-r)\sigma_A^2$, which implies (3).
Using (4) and (6), we get

$$C \approx k \int_{m}^{\infty} \lambda e^2 \exp(-\lambda e) \mathrm{d}e$$

Integrating, we get

$$C \approx k\left[\left(m+\frac{1}{\lambda}\right)^2 + \left(\frac{1}{\lambda}\right)^2\right] \exp(-\lambda m)$$

Using (8), $P(m) = 1 - \exp(-\lambda m)$ and for exponential distribution $\sigma_e = 1/\lambda$. Hence, replacing $\exp(-\lambda m)$ and λ, we get (7).

References

1. National Action Plan on Climate Change, GOI. http://www.moef.nic.in/downloads/home/Pg01-52.pdf. Last visited on 30 Sept 2016
2. India's Intended Nationally Determined Contribution: Working Towards Climate Justice. http://www4.unfccc.int/submissions/INDC/Published%20Documents/India/1/INDIA%20INDC%20TO%20UNFCCC.pdf. Last visited on 30 Sept 2016
3. Report of the Expert Group on 175 GW RE by 2022, NITI Aayog, GOI. http://niti.gov.in/writereaddata/files/writereaddata/files/document_publication/report-175-GW-RE.pdf. Last visited on 30 Sept 2016
4. Strategic plan for new and renewable energy sector for the period 2011–17, Ministry of New and Renewable Energy, Government of India, 2011
5. Das Abhik Kumar (2015) An analytical model for ratio based analysis of wind power ramp events. Sustain Energy Technol Assess 9:49–54
6. Mazumdar BM, Saquib M, Das AK (2014) An empirical model for ramp analysis of utility-scale solar PV power. Solar Energy 107:44–49
7. Kamath C (2010) Understanding wind ramp events through analysis of historical data. In: Transmission and distribution conference and exposition, 2010 IEEE PES in New Orleans, LA, United States, April 2010
8. Das AK, Majumder BM (2013) Statistical model for wind power based on ramp analysis. Int J Green Energy
9. Gallego C, Costa A, Cuerva A, Landberg L, Greaves B, Collins J (2013) A wavelet-based approach for large wind power ramp characterisation. Wind Energy 16(2):257–278

10. Bosavy A, Girad R, Kariniotakis G (2013) Forecasting ramps of wind power production with numerical weather prediction ensembles. Wind Energy 16(1):51–63
11. Kirby B, Milligan M (2008) An exemption of capacity and ramping impacts of wind energy on power systems. Electr J 2(7):30–42
12. Steffel SJ (2010) Distribution grid considerations for large scale solar and wind installations. In: IEEE, 1–3, transmission and distribution conference and exposition, 2010 IEEE PES
13. Das AK (2016) Forecasting and scheduling of wind and solar power generation in India. In: NTPC's 3rd international technology summit. Global Energy Technology Summit 2016
14. Indian Electricity Grid Code, Central Electricity Regulatory Commission, 2010. http://cercind.gov.in/2010/ORDER/February2010/IEGC_Review_Proposal.pdf. Last visited on 30 Sept 2016
15. Procedure for implementation of the mechanism of Renewable Regulatory Fund, Central Electricity Regulatory Commission. http://www.cercind.gov.in/Regulations/Detailed_Procedure_IEGC.pdf. Last visited on 30 Sept 2016
16. Framework on Forecasting, Scheduling and Imbalance Handling for Variable Renewable Energy Sources (Wind and Solar), Central Electricity Regulatory Commission. http://www.cercind.gov.in/2015/regulation/SOR7.pdf. Last visited on 30 Sept 2016
17. Model Regulations on Forecasting, Scheduling and Deviation Settlement of Wind and Solar Generating Stations at the State level. http://www.forumofregulators.gov.in/Data/study/MR.pdf. Last visited on 30 Sept 2016
18. Das AK (2014) Quantifying photovoltaic power variability using Lorenz curve. J Renew Sustain Energy, AIP 6(3):033124

Author Biography

Abhik Kumar Das holds a dual degree (B.Tech in Electronics and Electrical Communication Engineering and M.Tech in Automation and Computer Vision) from Indian Institute of Technology, Kharagpur, India. He has a vast experience in computational modelling of complex systems. He contributed in different verticals of analytical modelling related to renewable energy and techno-economics and published several well-cited research articles in internationally acknowledged journals and peer-reviewed conferences. He is a founding member of del2infinity, an accurate Wind Energy & Solar Energy Forecasting and Scheduling Solutions Company.

Smart Microgrids: Re-visioning Smart Grid and Smart City Development in India

Larisa Dobriansky, Girish Ghatikar and Dan Ton

Abstract This study addresses the role of Smart microgrids in shaping a "3.0 Smart Grid" to anchor Smart city development. The paper examines how "advanced or Smart microgrids" could contribute to developing an interactive, flexible, and innovative grid in India—one that would use information and communications technologies to increase the independence, flexibility, and intelligence for optimization of energy use and management within local energy networks and to cost-effectively integrate local energy resources into the Smart Grid. In this regard, the paper discusses integrating Smart microgrids with distribution utility "Advanced Distribution Management Systems," enabling "dynamic" microgrids that interact with distribution networks according to locally based Smart delivery architecture, with a view to harnessing cost-effectively the benefits of distributed resources for customers, the community, and the macrogrid. The paper also focuses on developing Smart microgrid "Infrastructure as a Service Platform" for resource-efficient community development, where microgrids manage and optimize local energy across multiple end-use sectors (power, transportation, water, waste, buildings, etc.). By leveraging data sets that span diverse facilities, systems, and purposes, Smart microgrids could interlink and optimize energy-using functions of diverse infrastructure systems and the built environment within cities. As part of this discussion, the paper will explore technical and regulatory innovations that could spur investment in advanced microgrids and the development of a 3.0 Smart Grid to help achieve India's Smart City policy objectives.

L. Dobriansky (✉)
Dobriansky Consultancy, Arlington, VA, USA
e-mail: Larisa.Dobriansky@gmail.com

G. Ghatikar
The India Smart Grid Forum (ISGF), New Delhi, India
e-mail: Ghatikar@gmail.com

D. Ton
Smart Grid R&D, U.S. Department of Energy, Washington D.C., USA
e-mail: Dan.Ton@hq.doe.gov

Keywords Smart microgrids · Smart grid · Smart cities · Advanced energy technologies · Regulatory reform · Integrated planning and operation

1 Introduction

India's vision for a Smart Grid interfaces integrally with its Smart City Mission objectives of achieving Smart energy management solutions (Smart metering, renewable sources of energy and efficient and green buildings), as well as achieving Smart solutions in such energy-intensive sectors as waste management, water management, and urban mobility.[1] India's "Smart Grid Vision and Roadmap" (Smart Grid Roadmap)[2] complements India's "Smart Cities Mission" and its objectives. These two blueprints share common goals and targets to advance clean, reliable, resilient, efficient, and economical prerogatives for electricity generation, delivery, and utilization. Developing the necessary synergies between these two strategies for action could galvanize continuous improvements in the quality of life, smarter solutions and increasingly cost-effective performance. Harmonizing the planning and implementation of these two strategies could efficiently capture the benefits of technology and lead to the efficient reorganization of India's electricity generation, transmission, and distribution system in a manner that further propels the transformation of its cities.

In fact, recognizing the sea change that is occurring within India's energy landscape, the Smart Grid roadmap represents a marked departure from the traditional, highly centralized and supply-push electricity sector strategy that India has pursued in the past. The Smart Grid roadmap sets significant targets at the distribution systems level to spur investment in distributed energy resources (DER) and to stimulate a customer/demand-driven strategy using Smart technologies.[3]

The Smart Grid roadmap seeks to achieve its targets by evolving a "Smart Grid," an electrical grid with automation, information, and communication technology (ICT)-based secure systems that can provide the capability for two-way power and information flows, monitor power flows from points of generation to points of consumption and control power flows or manage load to match generation in real time. Anticipating a rapid proliferation of distributed and renewable resources, this roadmap recognizes that it will be imperative to incorporate smarter automation, Smart meters, and information technology systems (IT) into the grid in order to manage onsite solar photovoltaic (PV), electric vehicles (EV), energy storage, and other distributed energy resources. Such IT systems can also enable utilities to

---

[1]Ministry of Urban Development (MOUD), Government of India (GOI), "Smart City Mission Statement and Guidelines," June 2015.
[2]Ministry of Power, Government of India (GOI), "Smart Grid Vision and Roadmap for India," August 2013. (India Smart Grid Vision and Roadmap.)
[3]India Smart Grid Vision and Roadmap.

better integrate intermittent renewable resources; provide increased visibility, predictability and event control of generation and demand (bringing flexibility to both generation and consumption); and enable utilities to reduce the cost of peak demand, reduce outages, and increase system efficiencies. Moreover, in the face of growing peak demand in India, the Smart Grid roadmap makes peak load management through load control, such as demand response (DR) and intelligent energy management, a high priority in efforts to secure reliable electricity services cost-effectively.[4]

India's Smart Grid vision seeks to overcome the linear constraints and limitations of the "centralized" legacy grid: centralized generation heat loss and transmission losses; demand growth stress; constraints to meet environmental objectives; infrastructure interdependencies; energy security vulnerabilities to natural and man-made disaster, with increased reliability costs; high cost to expand the development of new central plants and transmission lines to meet growing demand.[5]

Under this Smart Grid vision that embraces distributed resource control and heterogeneous service (in addition to centralized generation and delivery and universal energy access), Smart grid technologies could potentially expand the parameters of the legacy grid, evolving Smart distribution architecture, within which microgrid cells and distributed networked electricity systems could be nested. Evolving such Smart electricity architecture in synergy with the planning and implementation of India's Smart Cities Mission could produce smarter solutions, including outcomes that achieve much higher levels of energy performance (efficiency, power quality, grid resiliency and reliability, resource adequacy, customer choices, renewable energy integration and environmental quality); protect key community facilities and essential services during grid outages and energy disruptions; electricity access to over 240 million customers; and leverage private investment to help utilities, customers, and communities develop scalable and integrated energy systems [2].[6]

The main goal of this paper is to set out an "end-to-end" Smart electricity framework that could capture the synergies between India's Smart Cities and Smart Grid roadmaps. In particular, this paper provides tangible recommendations for deploying microgrids, as strategic technologies for increasing electricity access, resiliency, reliability, and sustainability in India. The "end-to-end" Smart electricity framework described in this paper addresses the following "building blocks" to modernize electrification, with a goal to maximize benefits and minimize costs to customers and grid operators from distributed resources:

---

[4]See generally, India Smart Grid Vision, and Roadmap.
[5]See Marnay [1].
[6]See, International Energy Agency (IEA), "India Energy Outlook," 2015 regarding energy access statistics.

Fig. 1 Interdependent relationships: Smart city, microgrid, and grid

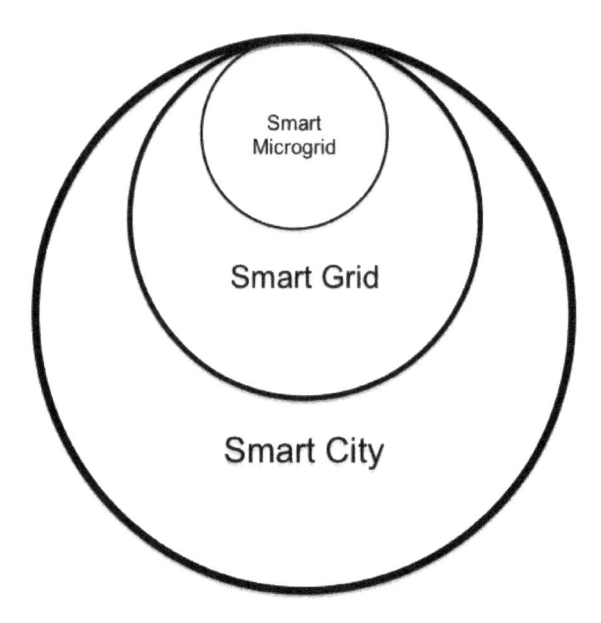

(1) Develop Smart microgrids, as essential catalysts for Smart city development to facilitate the convergence of energy supply and end-user sectors within community development and redevelopment using intelligent energy management.
(2) Expand and modernize electrification through developing a "3.0 Grid" in which Smart microgrids could maximize the net benefits of proliferating distributed energy resources (DER), for the grid, communities, and the market.
(3) Leverage standards for interconnectivity and interoperability between Smart microgrids and utility distribution systems to shape their interactions in support of diversifying electric resources; designing infrastructure that optimizes the management of energy requirements (heating, cooling, and power); and controlling and managing local reliability and increasing critical infrastructure resiliency.
(4) Develop Smart microgrid systems to foster the development of "integrated community energy systems," using advanced microgrid cells and networked cells to manage and optimize energy use across community end-user sectors for more efficient, secure, and sustainable energy usage.

This framework recognizes India's urgent need to meet the objectives of both its Smart Cities and Smart Grid visions and to maximize outcomes through the coordinated implementation of these visions. The framework described in this paper could leapfrog India in "clean" microgrid, Smart Grid, and Smart Cities innovation by accelerating the development of local renewable generation and efficient buildings within communities, while also supporting integrated regional sustainability and grid modernization strategies. Figure 1 illustrates these high-level interdependent relationships between Smart Cities, Smart Grids, and Smart microgrids and identifies Smart microgrids as a linchpin for India's vision for Smart cities and grid.

2 The Catalytic Role of Smart Microgrids

The US Department of Energy ("USDOE") and its national laboratories envision "advanced" or Smart microgrids playing an integral role in a modern electrical grid.[7] Equipped with new functionalities, enabled by advancements in information, communications, and control technologies (ICT, power electronics, modeling and simulation, distribution automation, data analytics), an advanced microgrid could serve as a "building block" for transforming a national electricity value chain, both upstream and downstream. USDOE and its national laboratories, along with other key agencies, are evaluating the changing capabilities of microgrids.[8]

The USDOE defines a "microgrid" as a "group of interconnected loads and distributed energy resources within clearly defined electrical boundaries that acts as a single controllable entity with respect to the grid and that can connect and disconnect from the grid to operate in grid-connected or island-mode."[9] India's "Model Smart Grid Regulations" define a Smart "microgrid" as "an intelligent electricity distribution system that interconnects loads, distributed energy resources and storage within clearly defined electrical boundaries to act as a single controllable entity with respect to the main (electric) grid. A microgrid uses information, communications, and control technologies to operate the system's distributed supply and demand resources in a controlled and coordinated way, either while connected to the main grid or while islanded; A microgrid can connect and disconnect from the grid to enable it to operate in both grid-connected or island-mode."[10]

---

[7]Sandia National Laboratories, Ton, US Department of Energy, et al, "The Advanced microgrid: Integration and Interoperability," SAND2014-1535, March 2014. The US Department of Energy first used the term, "Advanced microgrid" to distinguish the functionalities of these microgrids from today's conventional microgrids, sited and deployed as "niche" applications to serve the premium interests of the host customer (Sandia Advanced microgrid).

[8]Sandia National Laboratory Advanced microgrid, "Advanced microgrids" are dynamic in their functions, capable of interacting and interoperating with the macrogrid.

[9]Sandia National Laboratory et al, "The Advanced microgrid: Integration and Interoperability," SAND20141535, page 3 March 2014.

[10]Forum of Regulators, "Model Smart Grid Regulations," State Electricity Regulatory Commission (Smart Grid) Regulations, 2015. Also see, CIGRE C6.22 Working Group. In connection with its Rural Electrification Programs, the Indian Ministry of New and Renewable Energy (MNRE) has issued a proposal to define "microgrids" by size as a subset of "Minigrids." The term, "Minigrid," is used to represent the size of connected renewable power capacity of 10 kW or greater for both off-grid and grid-connected applications; the term, "microgrid," refers to connected renewable power capacity of less than 10 kW.

Advanced microgrids build upon "Smart grid"; such "advanced" or "Smart" (these terms shall be used in this paper interchangeably) systems incorporate intelligent energy management software and hardware for balancing energy supply and demand to maintain stable and reliable operations in real time. These intelligent or "Smart" electricity delivery networks manage and optimize multiple loads and distributed energy resources (together "DER," including distributed generators, storage devices, or controllable loads) in a controlled, coordinated way, either while connected to the main power network or while islanded. A "Smart" system contains ICT that enhances energy management and the optimization of the system's operations and components, including sensors, communications, and automatic control technologies used to generate, manage, distribute, and use electricity more intelligently and effectively.[11] In addition, the microgrid controller is a defining technology of microgrid systems that are required to manage the systems resources, as well as to connect and disconnect from the macrogrid.

Advanced or Smart microgrids contain the essential elements of the macrogrid; the ability to: Balance electrical demand with sources; Schedule the dispatch of resources; and Preserve grid reliability (both adequacy and security).[12] Key features of Smart microgrids include: ensuring maximum utilization of renewable energy sources and other assets; resource and load profiling, controlling and forecasting; load prioritization as critical or non-critical; real-time data acquisition and monitoring of electrical and physical signals; and minimization of outages and fast response to network disturbances through the automatic connect/disconnect of system components.[13]

The USDOE and its supporting agencies envision advanced microgrid systems that could contain multiple customers (or communities) and noncontiguous properties, multiple resources, resource interconnection on both sides of the meter, islanding capabilities, functionalities to provide grid services and use existing distribution infrastructure, as well as to create dedicated distribution infrastructure, including distributed networked electricity systems.[14] These community-scale advanced microgrid systems could be utility-owned, privately owned or have hybrid ownership and operating structures.

---

[11]Sandia National Laboratory, Advanced microgrid at 10, "Advanced hardware, intelligent inverters, Smart controllers and compatible communications will be the enabling technologies mix to maximize economics and operational benefits of advanced microgrid systems. Advanced and secure communication interfaces and Smart controls will increase the value of the energy provided by these advanced microgrid systems."

[12]Sandia National Laboratory, Advanced microgrid at 4.

[13]Sandia National Laboratory, Advanced microgrid Advanced microgrids operate using hierarchical levels of control to address system protection; automation; monitoring, optimization, resource scheduling and dispatching; and energy market, grid transactions. Also see, Electric Power Research Institute (EPRI), "Investigation of the Technical and Economic Feasibility of microgrid-Based Power System," (October 2008) regarding generally microgrid functions.

[14]Sandia National Laboratory, Advanced microgrid.

3 Developing an "Integrated Grid"

Both India's Smart City Mission and Smart Grid Vision and Roadmap recognize that the country's policy mandates, technology advancements, and changing consumer needs and interests will fundamentally affect the way in which energy is generated, delivered, and used. Increasing volumes of intermittent renewable energy in the generation mix, digitalization and higher penetrations of DER into the market, all necessitate grid modernization. A new grid operating system will be needed to meet these challenges, one that can manage bidirectional power and information flows to create an automated, widely distributed energy delivery network.[15] The "Grid 3.0" architecture is needed to achieve end-to-end integration and interoperability of many disparate systems and components, as well as to enable managing competitive transactions arising from a new restructured and distributed utility environment.[16] The 3.0 Grid will need to be interactive, flexible, and innovative, comprised of highly flexible, configurable, and interactive networks of utility, customer, and third-party applications; market data, price signals and transactions; "system of systems" operations for DER integration and load-side management.[17] The 3.0 grid operating system would be based on an "open-source design to facilitate the informational, financial, and physical transactions necessary to assure adequate security, quality, reliability, and availability of power systems operating in complex and continually evolving electricity markets".[18] This "integrated grid" will be able to take into account and value DERs in utility planning, investments, operations, and trading. The integrated grid architecture will support local customer and grid objectives, foundational for developing Smart microgrid communities. One such existing architecture being evaluated by the USDOE national laboratories is shown in Fig. 2.

Within an emerging distributed, heterogeneous communications and control network, the 3.0 Grid would support a new utility distribution system paradigm to respond to dynamically changing market conditions and manage customer-side resources, converting distribution system level functions from "conduit" to proactive in nature, with new Distribution System Operators (DSO) taking on transmission system-like functions to manage distribution planning, investments, and operations. These functions would entail: (1) Maintaining reliable distribution system operation with two-way, multi-point, reversible power flows arising from increasing volumes and diversity of distributed resources (voltage monitoring, telemetry, and real-time control); (2) Integrating and balancing distributed resources and load to shape load profile and peak demand and to enable multi-function DER to provide services to the bulk power system; (3) Achieving functional control of

---

[15]Electric Power Research Institute (EPRI), "Needed: A Grid Operating System to Facilitate Grid Transformation," July 2011 (EPRI 3.0 Grid).
[16]Electric Power Research Institute (EPRI), 3.0 Grid.
[17]See Footnote 16.
[18]Electric Power Research Institute (EPRI), 3.0 Grid at 6.

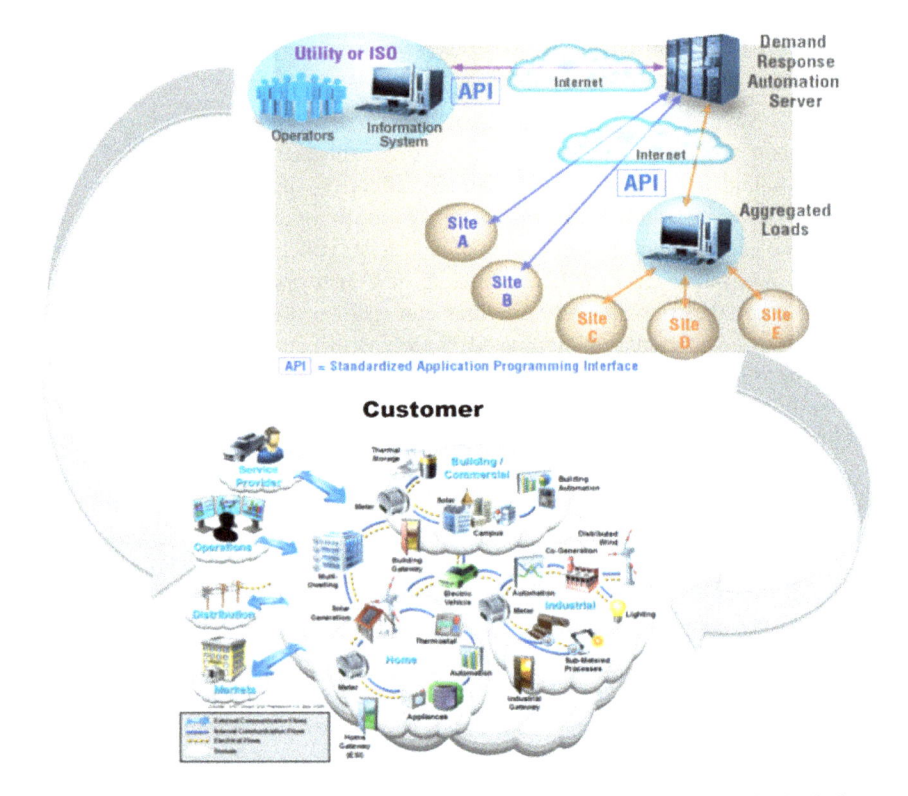

Fig. 2 Integrating microgrid resources with Smart grid domains using standardized platforms (*Figure sources* National Institute of Standards and Technology, and Lawrence Berkeley National Laboratory)

DER for real-time balancing and flexibility and services (such as reactive power and frequency control) to the local and bulk grid, modeling and forecasting load and DER growth; and (4) Defining and managing the transmission/distribution interface and addressing the changing nature of new resources and customers [3].

A 3.0 Grid would allow for the integration of energy sources and power delivery infrastructure at the local level and with the bulk power system. In this regard, Smart microgrids could play a vital role in the development of such local energy networks. Microgrid cells could become a connecting building block within a distributed networked electricity system to manage and optimize distributed resources.[19] Smart technologies would enable microgrid cells to be interconnected and to nest within a distributed networked electricity system that, in turn, would be connected to the bulk power system.[20] Such networking would allow for the sharing of generation,

---

[19]Electric Power Research Institute (EPRI), 3.0 Grid at 11–12.
[20]Id.

controllable load, and storage capabilities over wider areas.[21] These localized networks could address particular community and customer demands for cost control and managing price volatility, environmentally sustainable service, protection of critical infrastructure and municipal essential services, as well as support overall Smart community development, based on the local circumstances.[22]

If constituent microgrid design and operations are protocol compatible with the macrogrid and with neighboring microgrids, these systems could complement and participate as functioning units of a modernized grid.[23] These networks, consisting of modular facilities, could minimize stranded assets and assure that resources are used to their design capacity; maximize the value of individual technologies (electric vehicles, energy storage, demand response, distributed resources, as well as large central station renewables) from the nesting of microgrid systems with each other and with the bulk power system; increase the overall stability and security of delivery within the power system; reduce interdependencies between system components and increase resiliency to energy disturbances.[24] Overall, the 3.0 Grid would enable the seamless integration of such local energy networks with each other and with the Smart grid. Moreover, the networking of microgrid cells could provide a more cost-effective means of integrating large amounts of DER into the macrogrid than an elaborate utility command and control system, offering system efficiencies, reliability improvements, and more optimal management of dynamic sets of distributed and intermittent resources.[25]

Therefore, a 3.0 Grid would enable: Capturing the fuller potential of fusing advanced power and information technologies to increase the market value and use of distributed energy resources within microgrid systems and networked microgrids; Expanding electricity value chain parameters (integrating new resources and technologies, and accommodating new market players at the distribution system level); Facilitating competitive transactions resulting from competitive service offerings from utilities, third-parties, or prosumers (customer or third-party to grid, peer-to-peer transactions); and allowing for continuous improvement, innovation, and reconfiguration [4].

4 Leveraging the Benefits of Smart Microgrids

Microgrid "grid connection" means both interconnection physically and integration with the Smart Grid. Coordination and communication between a grid-connected microgrid and the grid are necessary to protect the integrity of electricity services.

---

[21]Id.
[22]See Footnote 16.
[23]Sandia National Laboratory, Advanced microgrid.
[24]See Footnote 16.
[25]Id.

The magnitude of energy use and operation of a microgrid in islanded (or autonomous) and grid-connected mode can have a major impact on overall grid-infrastructure planning and operations. The frequent and un-forecasted connect and disconnect functions of a microgrid without communications with the distribution utility or grid operators can jeopardize overall electricity system reliability, power quality, and cyber-security. An interactive microgrid can also be leveraged to cost-effectively harness the benefits of integrated distributed energy resources (DER) for grid services such as demand response. Such interconnected and integrated DER functions provide grid stability and enable grid service benefits by better integrating renewable generation through technology innovation and new market opportunities.

Advanced technologies enable interconnection and interoperability between Smart microgrids that are securely integrated with Smart Grid operators and service providers such as the transmission system operators, distribution system operators, and customers of the microgrid. Information and communication technologies play an important role to effectively utilize the net benefits of DER and their flexible energy use options offer significant opportunities for resource-efficient, and "integrated systems" approach to maximize microgrid benefits through local controllers and energy management systems. The various DERs such as distributed generation (e.g., energy storage, solar), electric vehicles, and loads can not only be optimally managed for local objectives—cost, energy, and carbon savings—they also can play a crucial role in integrated optimization for demand response. The two countries—United States and India—are jointly examining how ICT technologies and regulatory reforms could support the deployment of microgrids for Smart Grid and energy storage applications.[26]

4.1 Information and Communication Technology Platforms

Information and communication technologies (ICT) play a critical role in grid-connected microgrids. Previous experiences have shown benefits of enabling grid services and interconnection guidelines of microgrids [5].[27]

In addition to the interconnection guidelines (e.g., high/low frequency and voltage ride through) benefits for grid-connected and inverter-based DERs, open

---

[26]The US Department of Energy and the Indian Ministry of Science and Technology (Government of India) Federal Opportunity Announcement (FOA), USA—India Joint Clean Energy Research and Development Center: Smart Grid and Grid Storage Technology, FOA DE-FOA-0001606, CFDA Number: 81.087, July 2016.

[27]California Public Utilities Commission (CPUC), Electric Tariff Rule 21 (Rule 21), Interconnection requirements for inverter-based distributed energy resources (iDER), Decision 14-12-035, December 2014.

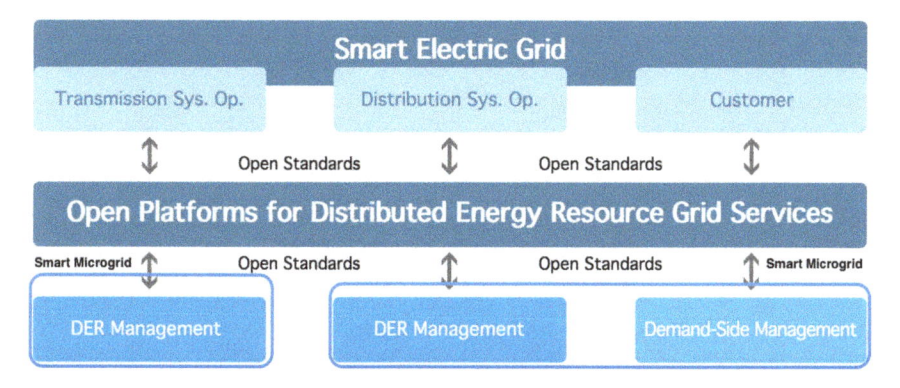

Fig. 3 Open platforms for end-to-end microgrid interoperability

standard-based ICTs enable interoperable and cyber-secure platforms to support integrated Smart microgrids with real-time measurement, monitoring, and control of microgrid systems. Such ICTs play a key role for customers and utilities to leverage DERs within Smart microgrids for autonomous operations and grid services. Customers can work with utilities and industries to enable technology innovation and new business opportunities that can integrate microgrid management systems across utility, system operators, and Smart City infrastructure to provide value-added services (e.g., provisioning of a microgrid for demand response programs to lower operational costs and improve grid reliability). Figure 3 shows the benefits of open standards for DERs within the microgrid to be integrated with the Smart electric grid domains—transmission, distribution, and customer network. While the new electricity markets are reviewing the DER and electricity market integration at the individual level (e.g., California's sub-metering and electricity rate tariffs for EVs), the minimum level of integration of DERs can be at the localized microgrid controller or the management system. The standardized ICT systems provide scalable two-way communication system within the microgrid for integrated DER management at the system level. The standardized utility advanced distribution management systems (ADMS) and DER management systems (DERMS) platforms can be connected to microgrids that need to be controlled with the local controller for efficient operation of a microgrid.

Figure 4 shows the existing standardized utility ADMS and DERMS infrastructure can also be integrated with new technology platforms for aggregated visibility and management of microgrids either though utilities or third-party networks without stranding the underlying assets.

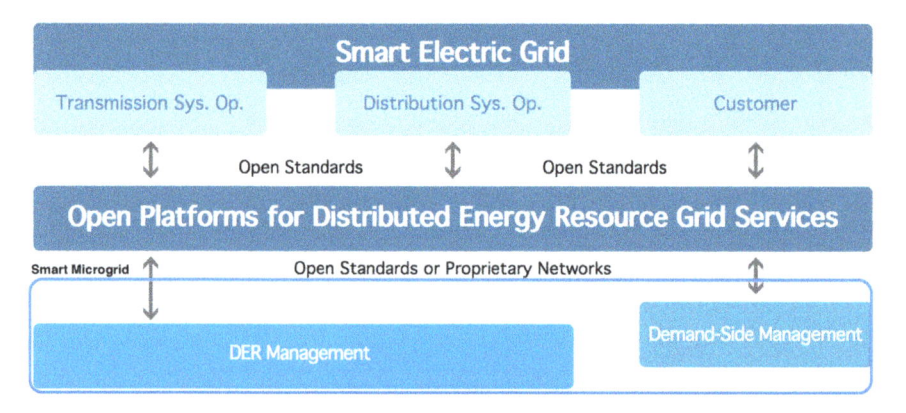

Fig. 4 Leveraged Smart microgrid platforms for distributed energy resource diversity

5 Platform for Intelligent Energy Management in Community Planning and Development

Interrelating Smart City, Smart Grid, and Smart microgrid strategies could maximize benefits to communities, customers, and the grid, while spurring technology innovation and competition. Coordinating these strategies could provide cities a new discipline for intelligent energy management with respect to their Smart, sustainable, and secure community planning and development. Development opportunities would be broadened, spurring a range of Smart city applications and services and generating new revenue streams. Advanced microgrids and distributed networked electricity systems could serve as design and operational tools for communities, customers, and the grid.[28] Cloud-based technology could enable networked microgrids to interconnect community users with energy using infrastructure and facilitate interactions though an "Internet of Things" (IoT).[29] Moreover, within this context, Smart microgrids and networked microgrids could catalyze both Smart grid and Smart city evolution.

As aforementioned, advanced microgrids have the ability to cluster loads and DER units as an integrated system, which operates in islanded or parallel-grid modes (microgrids are generally connected to the larger utility through a point of common coupling), and which could contribute to Smart distribution through faster control of many individual DER units. In this way, Smart microgrids could help communities to evolve integrated energy solutions in developing their infrastructure and built environment. Using ICT capabilities and a resource-efficient systems approach, Smart microgrids could interrelate energy generation and consumption through clustering compatible uses of infrastructure, buildings, and public works,

---

[28]Dobriansky, Revisioning Smart Community Development.

[29]Dobriansky, Revisioning Smart Community Development; Ghatikar [6].

building local networks to advance resource integration and convergence (thermal and electric, water, waste, transport, buildings, etc.), increase efficiencies and optimize investment and energy usage.[30]

Cities could combine, within their land use development and growth management processes, the design and development of Smart local delivery systems with Smart growth design, Smart grid and the IoT. Combining these capabilities into land use planning and development will enable communities to achieve intelligent load and energy resource management, balancing loads with variable renewable energy using efficiency and demand response capabilities; diversify sources; integrate storage, load shifting and prioritization, base plus variable generation with Smart grid technologies; increase resiliency and provide self-healing in the event of disruptions; and amplify sustainability, while facilitating orderly, capital efficient and environmentally sound application of distributed resources.

Microgrid configurations could be localized to customer and community circumstances and needs, but also operates under compatible standards for interaction with and providing services to (energy, capacity, ancillary services) the macrogrid and power markets. Smart microgrid infrastructure (microgrids and Smart networks) could serve as a "Platform" for applying intelligent energy management at all services levels: Phase I Individual Service Level (individual city operations); Phase II Vertical Service Level (integration of related processes and services by Smart technology) and Phase III Horizontal Service Level (seamless integration of different service areas and end uses within an efficient Smart city ecosystem) [8]. This "Infrastructure as a Service Platform" would leverage data sets that span diverse facilities, systems, and purposes to interlink and optimize energy using functions of diverse infrastructure systems and the built environment within communities. Local energy networks could promote consumer engagement with resources to solve issues locally; increase the deployment of local renewables in residential, commercial and industrial energy usage; facilitate two-way power and information flows in distribution; and shift consumers from passive behavior to active control and management of their energy consumption and generation.[31]

Finally, developing and deploying Smart microgrid infrastructure could also benefit community planning and development processes in the following significant ways: (1) Understanding the impacts of embedded energy costs and operational energy needs of infrastructure systems and urbanization; Assessing and quantifying the benefits and costs of alternative technologies, practices and development scenarios; and Developing cost-effective decision support tools and methods for community energy systems planning; (2) Incorporating energy supply and demand infrastructure analyses of alternative energy and resource development options into housing, land use, water supply, and wastewater, transportation, waste recycling and reuse and other municipal processes; and (3) Building energy surety and resiliency as a bedrock for sustainability, while harnessing sustainability to mitigate

---

[30]See, for more detail, Dobriansky [7].
[31]Mohn, Smart microgrids.

impacts, diversify supplies, harden critical infrastructure, increase resiliency, stabilize energy costs and reduce reliance on fossil fuels. Evolving Smart microgrid infrastructure is a means for reducing risks and vulnerabilities, developing the fuller potential of the IoT, Smart Grid, and Smart Growth, and maximizing outcomes by capturing co-benefits.

6 Regulatory and Market Innovations

Development of a new policy "ecosystem" will be needed to support investment in a 3.0 Grid and the Smart microgrid infrastructure that can advance the objectives of India's Smart City Mission. This policy ecosystem would be designed to capture the benefits of DER and the value that Smart microgrids create through intelligent and resource-efficient management of DER as part of both utility and community planning, investment, operations, and market trading. In the case of the utility regulatory regime, this would entail supporting in the following ways a new macrogrid design, new resource valuation methods and changes in economic regulation and market rules: (1) **Smart Grid Architectural Design** to advance interoperability and integration "end to end" within the power value chain in order to standardize the use of microgrid systems and DER as part of power system planning, grid operations, and power market trading, with a view to creating highly flexible, configurable, and interactive electricity networks; (2) **New Resource Valuation Methods** that can consistently measure the costs and benefits of DER and Smart microgrid systems so that utilities can evaluate the cost-effectiveness of these resources compared to traditional options in their planning, procurement, and investment decision-making[32]; (3) **New Utility Regulatory Compact** that will align utility financial interests with the creation of long-term customer value; that will change the incentives of traditional utility rulemaking, including cost of recovery approaches, and rate design, to make utility decision-making indifferent to ownership and focused on achieving the most cost-effective solutions; that establish Distribution System Operators to respond to dynamically changing market conditions, and to manage distribution planning, investments and operations; and that move to more cost-reflective pricing; and (4) **New Power Market Rules and Products** that value resources based on timing, location, flexibility, predictability, and controllability.[33]

In addressing Smart microgrids and Networking of microgrids, the policy "ecosystem" needs to: (1) Establish a consistent definition of microgrids so their benefits can be taken into account in utility and community planning, investment

---

[32]EPRI, "The Integrated Grid: Realizing the Full Value of Central and Distributed Energy Resources," 2014.

[33]See, for example, New York State Department of Public Service Proceeding on Motion of Public Service Commission in Regard to Reforming the Energy Vision. CASE 14-M-0101, "Developing the REV Market in New York: DPS Staff Straw Proposal," August, 2014; See also, Dobriansky Smart City Tutorial.

decision-making and operations; (2) Establish standards that set specifications and requirements for microgrid controllers; interconnection standards that take into account the functions and benefits of microgrid systems; and protocols and implementation guidelines for communication, interconnectivity and interoperability between microgrid and utility systems; (3) Develop valuation methodologies and cost/benefit frameworks that can consistently assess, measure, and monetize the costs and benefits associated with Smart microgrid management and optimization of DER.

Communities will need to: (1) Develop consistent and verifiable methods to assess the benefits and costs of Smart microgrid systems in their consideration of alternative technologies, practices, and development scenarios; (2) Develop cost-effective decision support tools and methods for integrated community energy systems planning and development; (3) Evaluate and determine governance structures and procurement policies and procedures that take into account the net benefits of Smart microgrid infrastructure to achieving Smart city objectives and that can leverage private investment; and (4) Develop coordinated planning and development processes across organizational divisions to address the challenge that new Smart grid and Smart microgrid architecture will necessitate a certain level of openness to connect and integrate historically resource silos, while protecting end-user privacy rights and mitigating cyber-security risks.

7 Conclusions and Recommendations

The "end-to-end" Smart electricity framework described in this paper could provide India a means for achieving cost-effectively the objectives of its Smart Grid and Smart Cities visions by coordinating these programs using Smart microgrid deployment strategies, with a view to "leapfrogging" clean, secure, efficient, and reliable innovation in India.

The design of open innovation of Smart microgrids, in support of India's Smart City and Smart Grid visions should be based on guidelines for the ICTs and cyber-security that can spur new competitive market opportunities through digital infrastructure and provide customers with choice in efficient and cost-effective operation of microgrids. In addition, the interconnection guidelines for inverter-based DERs that are part of the microgrids can play a key role in enabling grid reliability and higher penetration of renewable generation.

To reap the benefits of microgrid technologies and to capture synergies and co-benefits, India's Smart City and Smart Grid strategies need to be coordinated and harmonized; and mechanisms need to be established for implementing coordination between relevant governmental agencies and stakeholders.

In light of the objectives of the Smart City Mission and the Smart Grid Vision and Roadmap, as well as the significant renewable energy, DER, and energy access targets that India has set, it would be advisable to evaluate the need to develop a 3.0 Grid, as well as to design and develop demonstrations to measure and quantify the cost-effectiveness of alternatives, such as microgrid systems, to traditional utility investments.

To provide an economically viable opportunity for Smart microgrid development, India will need to establish a consistent definition of microgrids; valuation methods and cost/benefit analytical frameworks that can quantify and monetize the value that microgrid energy management systems can create, as distinctive from technology-specific valuations.

Utility regulatory reforms are needed to change utility incentives and to align utility financial interests and engagements for the creation of long-term customer value.

References

1. Marnay C (2010) Microgrid architecture. In: 4th international conference on integration of renewable and distributed energy resources, Dec 2010
2. Dobriansky L (2016) The smart microgrid solution: rethinking and revisioning smart community development. In: 7th world renewable energy technology congress, Aug 2016 (Dobriansky, Revisioning Smart Community Development)
3. Erickson et al (2015) Distribution system planning and innovation for distributed energy futures. Power sector. Springer International Publishing, Switzerland, Aug 2015
4. Dobriansky L (2016) Grid 3.0: new parameters, players and structures. In: 2016 OATI conference, Oct 2016
5. Ghatikar G, Mashayekh S, Stadler M, Yin R, Liu Z (2015) Distributed energy systems integration and demand optimization for autonomous operations and electric grid transactions. Appl Energy. http://dx.doi.org/10.1016/j.apenergy.2015.10.117
6. Ghatikar G (2017) Internet of things and smart grid standardization. In: Geng H (ed) Chapter 30 in the internet of things and data analytics handbook, First edn. Wiley, USA. ISBN: 978-1-119-17364-9
7. Dobriansky L (2017) Smart microgrids, smart grid and smart city development in India. 2017 India smart grid week tutorial 2, "Leading transition to a smart city." Mar 2017 (Dobriansky Smart City Tutorial)
8. Mohn T (2016) "21st century grid—smart microgrids," ISGW 2016. This PPT addresses the role of microgrids in developing smart, sustainable and secure communities (Mohn, Smart Microgrids)